Rudolph Bergh

# Scientific Results of the Exploration of Alaska by the Parties Under the Charge of W.H. Dall,

during the years 1865-1874. On the nudibranchiate Gasterpod Mollusca of the North Pacific Ocean: with special reference to those of Alaska: second part

Rudolph Bergh

**Scientific Results of the Exploration of Alaska by the Parties Under the Charge of W.H. Dall,**
*during the years 1865-1874. On the nudibranchiate Gasterpod Mollusca of the North Pacific Ocean: with special reference to those of Alaska: second part*

ISBN/EAN: 9783337319441

Printed in Europe, USA, Canada, Australia, Japan

Cover: Foto ©Andreas Hilbeck / pixelio.de

More available books at **www.hansebooks.com**

# SCIENTIFIC RESULTS

OF THE

# EXPLORATION OF ALASKA.

BY THE PARTIES UNDER THE CHARGE OF W. H. DALL,

DURING THE YEARS 1865-1874

## VOL. I.

ARTICLE VI. On the Nudibranchiate Gasteropod Mollusca of the North Pacific Ocean, with special reference to those of Alaska. Second Part.

BY DR. RUDOLPH BERGH, COPENHAGEN.

WASHINGTON, D. C.

JANUARY, 1880.

# ARTICLE VI.

## ON THE NUDIBRANCHIATE GASTEROPOD MOLLUSCA OF THE NORTH PACIFIC OCEAN, WITH SPECIAL REFERENCE TO THOSE OF ALASKA.

### BY DR. R. BERGH, COPENHAGEN.

## PART II.

### DIAULULA, Bgh.

*Diaulula*, Bgh., Malacolog. Unters. (Semper, Philipp. II, ii), Heft xiii, 1878, p. 567; Heft xiv, 1878, p. xxxv. Gattungen nordischer Doriden, Arch. f. Naturg., xxxv, 1, 1879, p. 343.

Forma corporis sublepressa. Dorsum minutissime villosum, holosericeum, molle. Tentacula digitiformia. Apertura branchialis rotundata, crenulata; folia branchialia tripinnata. Podarium antice bilabiatum, labio superiore medio fisso.

Armatura labialis nulla. Lingua rhachide nuda, pleuris multidentatis, dentibus hamatis. Prostata magna; penis inermis.

In their general form the *Diaulula*[1] somewhat resemble the *Discodorides* and the *Thordisa*,[2] although their habitus still is peculiar. The back is villous, as in these genera and especially as in the *Thordisa*, but finer and more velvet-like. The tentacles are finger-shaped, smaller than in the *Discodorides*, larger than in the *Thordisa*. The branchial-slit is rounded, crenulated; the branchial leaves tripinnate. The anterior margin of the foot bilobed, the upper lip broader, with a median fissure. As in the *Thordisa*, there is no armature of the lip-disk. The radula nearly agrees with that of the *Discodorides*; the rhachis is naked; on the pleurae there is a rather broad series of plates of the usual hook-shape. The stomach is enclosed in the liver (not free, as in the *Discodorides* and in the *Thordisa*). As in the *Discodorides*, there is a large prostate and an unarmed penis.

Only the following species appears to be hitherto known, from the northern Pacific.

1. *D. Sandiegensis* (Cooper).

---

[1] *Diaulus*, medicus, cf. Martialis, I, 48, p. 40.
[2] Cf. my Malacolog. Untersuch. (Semper, Philipp. II, ii), Heft xii, 1877, p. 518, (*Discodoris*); p. 540 (*Thordisa*).

**1. D. Sandiegensis,** Cooper. Plate V, fig. 3-9.

*Doris (Actinocyclus?) Sandiegensis,* Cooper, Proc. of the California Acad. of Nat. Sciences, ii (1862). 1863, p. 204;[1] iii (1863); 1868, p. 58.

Color corporis e brunneo lutescens, annulis nigris maculatus; vel brunneus.

*Habitat.* Oceanum Pacificum orient. (San Diego Bay; Santa Barbara; Sitka Harbor; Puget Sound).

According to Cooper, numerous specimens of this species were found from November to May among grass on mud flats in San Diego Bay, at or near low water mark; according to Cooper, it is a very "active" species; Cooper later obtained two specimens at Santa Barbara Island, on rocks at low water. During the expedition to Alaska a specimen was taken by Dall in Sitka Harbor, on algæ, in August. 1865, at the depth of six fathoms (another in August, 1873, in Puget Sound, by Dr. Kennerly, on algæ, at low water).

Through the kindness of Dall, I have seen the original (rather rough) drawings of this species by Cooper; a colored one represents the back bright chocolate-brown, with six black rings, of which there are two smaller ones between the rhinophoria; the rhinophoria, the gill and the foot seem bright-yellowish; one figure shows five, another six branchial leaves.

The length of the first specimen, sent to me preserved in spirits. was about 22.0 mm., the height reaching 9.0 mm., and the breadth 13.0 mm.; the breadth of the foot reached 10.0 mm., the height of the rhinophoria 2.0 mm., the branchial leaves 3.3 mm. The color was uniformly brownish-gray; nearly symmetrically on each side of the true back was an annular black spot.

The form of the rather soft body elongate-oval, not much depressed. The head quite concealed between the mantle and the foot; the outer mouth had the form of a vertical slit; at each side a short finger-shaped tentacle. The margin of the rather large rhinophor-holes rather prominent, crenulate; the rhinophoria strong, the club

---

[1] "Pale brownish-yellow, with large, annular, brown spots, irregularly scattered, varying from twelve to twenty, or entirely brown. Surface slightly rough; sometimes a little tuberculated. Dorsal tentacles conical, retractile; branchiæ large, rising in five parts, which become tripinnately divided, expanding so as to cover the posterior third of the body like an umbrella. Mouth proboscidiform, with two short lateral tentacles. Length, 3½ inches; breadth, 2½ inches; height, ½ inch.—COOPER, l. c.

with about thirty leaves (on each side). The back all over minutely and densely villous (fig. 3). The margin of the rather wide (5.0 mm.), roundish branchial aperture like the margin of the rhinophor-holes, prominent, finely crenulate; the branchial leaves (retracted) six in number, very strong, tri- or quadripinnate. The anus strong, about 1.5 mm. high, cylindrical, closing the branchial ring posteriorly; the renal pore as usual. The edge of the mantle rather thick, projecting about 2.0 mm. from the body; the sides low. The genital opening as usual, with two distinct apertures at the bottom. The foot strong, broad, somewhat narrower towards both ends; in the anterior margin a strong furrow, towards the median line deeper and forming two lips; the superior broader and divided in the median line.

The cerebro-visceral ganglia kidney-shaped, the visceral larger than the cerebral; the pedal of roundish contour, scarcely larger than the visceral. The buccal ganglia of oval form, connected by a short commissure; the gastro-œsophageal roundish, short-stalked, in size about one-fifth of the former, with one very large and one large cell.

The eyes short-stalked, with black pigment and yellowish lens. The otocysts scarcely smaller than the eyes, overcrowded with otokonia of the usual kind. The leaves of the rhinophoria strengthened with long, perpendicular spicula, calcified at the surface. The tentacula with a mass of shorter, but otherwise similar spicules, lying irregularly. The villi of the back closely set with perpendicular spicula (fig. 3). The anal papilla with long, perpendicular spicules; the stalk of the branchial leaves with many shorter spicula, irregularly situated; in the leaves themselves were no spicules. In the interstitial connective tissue large spicules were seen rather sparsely.

The oral tube was about 1.5 mm. long, wide, with strong longitudinal folds. The bulbus pharyngeus only about 4.0 mm. long, by a height of 2.0 mm., and a breadth of 4.0 mm.; the rasp-sheath very prominent on the hinder part of the under side of the bulbus; the inner mouth with a yellowish, not thin, cuticula. The tongue with nine rows of teeth, in the rasp-sheath also eleven rows of developed and two of not quite developed teeth, the total number thus being twenty-two. In the posterior rows of the tongue the number of plates was twenty-eight or twenty-nine, on each side, and seemed in the succeeding rows not to surpass thirty. The color of the teeth horn-yellowish; the height of the outermost 0.06 to 0.08 mm., the height rising to about 0.18 mm. The form of the teeth as usual; the wing rather narrow; the innermost (fig. 5*aa*, *b*) not very different from the others (fig. 5, 6), the body of the outermost three or four (fig. 4aa, 7), as usual, of reduced size.

The glandulæ salivales (5.0 or 6.0 mm.) long, in the anterior part about one-third larger than in the rest, measuring 1.0 mm. in diameter, yellowish; in the rest of the length much narrower, whitish.

The œsophagus is about 9.0 mm. long, rather wide. The stomach is included in the liver, not spacious. The intestine appears on the surface of the liver in the usual manner, passing forwards, forming a short flexure, and running straight backwards to the anal tube, which has in its interior many fine longitudinal folds; the total length of the intestine about 20.0 mm., with fine longitudinal folds through its whole length. The cavity was empty. The liver yellowish, about 17.0 mm. long, by a breadth of 8.0 mm., and a height of about 6.0 or 7.0 mm.; the anterior end truncate, the posterior end rounded; on the right side of the forepart a flattened impression for the anterior genital mass. The vesica fellea, as usual, behind and at the left side of the pylorus, elongate-pyriform, grayish, taken together with its duct about 2.5 mm. in length.

The heart as usual. The two gland. sanguineæ as usual, whitish; the foremost more triangular, about 3.5 mm. long; the posterior broader, about 2.0 mm. long.

The gland. hermaphrodisiaca with a rather thick yellow layer clothing the greater part of the surface of the liver (except the posterior end); in the lobules of the organ were rather large oögene cells and masses of zoösperms. The anterior genital mass large, compressed, about 10.0 mm. long, by a height of 6.3 mm., and a breadth of 3.0 mm. The ampulla of the hermaphroditic duct strong, grayish, when unrolled about 25·0 mm. long, somewhat coiled on the anterior end of the left side of the mass and on its inferior flattened edge behind the large prostate; it reaches a diameter of 1.2 mm. The male branch of the ampulla (fig. 8a) thin, white, passing into the narrow inferior end of the prostate, thus forms the fore-end of the whole genital mass. The prostate (fig. 8b) is of dirty yellow color, flattened and irregularly pyriform, the length about 6.3 mm., by a breadth of as much as 3.0 mm.; the spermatoduct (fig. 8c) issuing from the upper part of the posterior side of the gland, in its first thicker part nearly as long as the prostate; in the rest of its length thinner, making several coils and passing (fig. 9a) into the male organ. The retracted penis (fig. 8d) strong, about 2.5 mm. long, the præputium with fine longitudinal folds (fig. 9), from the uperture upwards and nearly filled by the glans, which had nearly the form of a human penis, with a well developed head with round aperture; this head seemed covered with very small, low and rounded, soft papillæ. The spermatotheca were whitish, spherical, of the diameter of about 2.3 mm., filled with epithelium, fatty matter and altered semen; the chief duct a little longer than the spermatotheca, gradually passing into the simple vagina, that was about half as much in length (and was filled with sperma). The spermatocysta of violet-gray color, somewhat flattened, of oval outline, of the length of about 2.3 mm., filled with sperma. The posterior half, or a little less, of the large mucous and albuminous gland, chalk-white; the anterior, more than half, of grayish or (on the left side) yellowish color; the structure as usual.

A variety of the species (according to Dall, it also belongs to this species) was, moreover, obtained by Dr. Kennerly, in August, 1873, on algæ, at low water, in Puget Sound, Washington Territory (fig. 6–9).

The single individual was rather large; the length 40.0 mm., by a breadth of 28.0 mm., and a height of 13.0 mm.; the breadth of the foot 15.0 mm., of the margin of the mantle 11.0 mm.; the height of the rhinophoria 5.0 mm., of the branchial leaves nearly 5.0 mm. The color of the upper side obscure olive-gray, with rather large (diameter about 4.0 mm.) black and blackish spots; the under side yellowish. The general form and the head, with the tentacles, as above described. The openings of the rhinophor-holes as above, the club with about twenty five leaves. The branchial opening as above (diameter, 3.5 mm.); the retracted branchial leaves six in number; the anal tube nearly 3.0 mm. high. The back villous, as in the typical individual. The foot as above.

The peritoneum colorless, without larger spicula; but in the region of the ventricle of the heart the pericardium is brownish.

The central nervous system as above; the proximal olfactory ganglia bulbiform, a little larger than the buccal; the distal ones smaller than the proximal, at the root of the club of the rhinophoria. The buccal ganglia of oval form; the commissure between them being about one-third of the largest diameter of the ganglia. The eyes, the otocysts, the leaves of the rhinophoria and the villi of the back as above.

The oral tube large, of a length and diameter of 4.0 mm. The bulbus pharyngeus 4.0 mm. long, by a height of 4 and a breadth of 3.5 mm.; the sheath of the radula less prominent than in the former specimen; the cuticula of the lip disk as above. The tongue with ten rows of plates, further back eleven developed and two younger rows, the total number thus twenty-three. In the posterior rows of the tongue there were as many as thirty-four dental plates on each side of the rhachis; they resembled those above described (fig. 6, 7).

The salivary glands yellowish, ribbon-shaped. The stomach as above. The anteriorly proceeding part of the intestine 7.0 mm. long, by a diameter of about 2.0 mm.; the receding part about 20.0 mm. long, by a diameter of 1.5 mm. In the stomach and the rectum were pieces of a Keratospongia and different Diatomacea. The liver 23.0 mm. long, by a breadth and a height of 11.0 mm.; the anterior end truncate, with a median deep and narrow slit for the œsophagus and for the intestine; the right anterior half of the liver rather excavated, especially beneath; the substance of the liver yellow.

The foremost glandula sanguinea about 4.5 mm. long, by a breadth of 2.5; the posterior 4.0 mm. long, by a breadth of 2.5 mm.; both very flattened (about 0.8 mm. thick), grayish-yellow. The kidney with its whitish network, contrasting prettily with the yolk-yellow hermaphroditic gland; the urinary chamber not wide; the tube on its floor thin.

The hermaphroditic gland clothing nearly the whole liver (with its posterior end), as in the former specimen. The anterior genital mass about 11.5 mm. long, by a height of 9.5 and a breadth of 5.0 mm., the ducts also projecting 3.0 mm. The ampulla of the hermaphroditic duct yellowish-white, about 35.0 mm. long, by a diameter of 1.25 mm., running across the upper part of the left side of the genital mass, and forming several windings on the anterior part of the upper margin.

The large prostate as above (fig. 8b), dirty yellow; 7.5 mm. long, by a diameter at the upper end of about 4.0 mm.; the part (fig. 8c), from which the spermatoduct proceeds, much brighter than the rest of the organ. The thin spermatoduct forming (fig. 8) a little coil at the upper end of the penis; when unrolled about 12 mm. long. This last (fig. 8d, 9) organ strong, about 4.0 mm. long, by a diameter of 1.5 mm.; the prominent orifice in the vestibulûm (fig. 8e) with strong longitudinal folds; the glans conical, filling nearly half (fig. 9) of the cavity of the organ, the surface (under a power of 350) smooth. The spermatotheca whitish, spherical, with a diameter of 3.5 mm.; the spermatocysta short, sausage-shaped, about 4.0 mm. long, of reddish-yellow color. The duct from the spermatotheca to the vagina rather thick, 3.5 mm. long; the vagina larger than the penis, 6.0 mm. long, by a diameter of 2.5; the inside with fine longitudinal folds, and with sperma in the cavity. The mucous gland large, 9.0 mm. long, by a height of 7.5 and a thickness of 4.0 mm.; whitish, yellowish chalk-white and yolk-yellow; the duct rather short, with the usual strong fold. The vestibulum with longitudinal folds.

### JORUNNA, Bergh.

*Jorunna*, Bgh., Malacolog. Unters. (Semper, Philipp. II, ii) Heft x, 1876, p. 414, note. Gatt. nord. Doriden, Arch. für Naturges., xxxv, i, 1879, p. 346.

Corpus subdepressum; dorsum minutissime granulatum, sub-asperum, branchia e foliis tripinnatis formata; tentacula digitiformia; podarium sat latum, margine anteriore sulcatum, labio superiore latiore et medio fisso.

Armatura labialis nulla. Radula rhachide nuda, pleuris multidentatis, dentibus hamatis. Penis stylo armatus; glandula et hasta amatoria.

This genus was established by the author on the *D. Johnstoni* (1876) in reference to the results of the anatomical examination of Hancock and Embleton; he regarded it as nearly allied to the *Kentrodorides*, just founded by him.[1] After the present examination of the *D. Johnstoni* by the author he is not entirely certain of a generic difference between the *Jorunnæ*[2] and the *Kentrodorides*. The latter have been examined only from rather insufficient material, and the hasta has not been seen in any of the species, only a papilla in connection with a peculiar gland; still the *Kentrodorides* are of a quite different habitus, very soft, and the upper lip of the anterior margin of the foot is more developed, while the innermost plate of the tongue is somewhat different from the others. If not identical with the *Kentrodorides*, the *Jorunnæ* are certainly very nearly allied to them.

The *Jorunnæ* are rather depressed; the back finely granulated, covered with equal minute papillulæ; the retractile gill formed of tripinnate leaves; the tentacles digitiform; the foot rather broad, deeply grooved in the front margin, and the upper lip of this larger and cleft in the middle line. The lip-disk not armed, covered with a simple cuticula. The rhachis of the radula naked, the pleuræ with many hook-formed plates. In the vestibulum genitale are four apertures:

[1] R. Bergh, Malacolog. Unters. (Semper, Philipp. II, ii) Heft x, 1876, p. 413-427, Tab. XLIX LI.
[2] Jorunna, Björnis filia. Laxdäla Saga. Hafniæ, 1826, p. 21.

one for the penis, which is armed with a stylus; another for a *hasta amatoria*, through which opens a peculiar gland (quite as in the genus *Asteronotus*);[1] a third for the vagina, and the fourth for the duct of the mucous gland.

Only one species of the genus seems hitherto known, belonging to the northern part of the Atlantic Ocean. The spawn of the species is known from Alder and Hancock, but nothing else is known of the biology of the animal.

1. *J. Johnstoni* (A. et H.).
   *Doris Johnstoni*, A. et H. Oceanum Atlantic. septentr.

Jorunna Johnstoni (A. et H.). Plate VIII, fig. 19; Plate IX, fig. 1-11.

*Doris Johnstoni*, Alder et Hanc. Monogr. Part I, 1845, fam. 1, Pl. 5; Part V, 1851, fam. 1, Pl. 2. f. 8-11.

*Doris Johnstoni*, Hanc. et Embleton, Anat. of Doris. Philos. Trans. 1852, II, p. 212, 215, 216, 220, 233, Pl. XII, f. 2, 10; Pl. XIV, f. 9, 10; Pl. XV, f. 1-2; pl. XVII, f. 2-3.

*Doris Johnstoni*, Forbes and Hanley, Hist. of Br. Moll., III, 1853, p. 564.

? *Doris tomentosa*, Cuv., Fischer. Journ. de Conchyl., 3me Sér., x, 1870, p. 290-293; XV, 1875, p. 211, note.

? *Doris tomentosa*, C. Verany, catalogo. 1846, p. 16-21. Ver., Hanc. et Embleton, l. c. 1852, p. 220.[2]

? *Doris tomentosa*, C. Philippi, En. Moll. Sic. I., 183, p. 104; II, 1844. p. 79, Tab. XIX, f. 9.

Color flavescens, dorso interdum maculis fuscis seriatis ornatus; rhinophoria fusco-maculata; branchia albescens.

*Hab.* Oceanum Atlanticum septentr.

This species, that was first described by Johnston under the name of *D. obvelata* (Müller), was (1845) established by Alder and Hancock. Hancock gave a series of anatomical remarks upon this very interesting form and of figures referable to it. Since then nothing new seems to have been published about the species; but a few months ago I (l. c.) gave a short notice of the generic characters of the group.

Of this form I have only examined a single specimen, captured in March, 1870, in the neighborhood of Hellebäk, on the north coast of Seeland (Denmark).

---

[1] R. Bergh. Ueber das Geschlecht *Asteronotus*, Ehrbg. Jahrb. der Deutschen Malakozool. Ges., iv, 1877, p. 161-173, Taf. I-II.

[2] According to Hancock and Embleton (l. c., p. 220), the dart (hasta amatoria) in *Doris Johnstoni* is straight, in *D. tomentosa*, Ver., curved.

The specimen was of a uniform yellowish color; the rhinophoria finely dotted with brown (but not the branchial leaves). The length of the rather contracted and somewhat contorted individual was about 18.0 mm. by a greatest breadth of 10.0 and a height of about 7.0 mm.; the height of the (retracted) rhinophoria 2.5, of the tentacles nearly 1.5, of the (retracted) gill 2.5 mm.; the greatest breadth of the mantle-margin 3.5 mm., of the foot 5.0 mm.

The form is elongate-oval, the mantle-margin rather thick, not very broad. The back covered all over with very minute granules, sometimes, especially on the middle of the back, crowded in irregular and roundish small groups; the under side of the mantle-margin smooth. The (contracted) openings of the rhinophor-holes appear as a simple transverse slit, the granules of the back reaching forward to the opening, those in this neighborhood not larger than the rest. The club of the rhinophoria stout, with about thirty[1] broad leaves. The opening of the gill-cavity small, transverse, triangular-crescentic, with the convexity forwards (as contracted); the granules of the back reaching to the very margin of the gill-slit, but not larger than the rest. The gill consisting of eleven branchial leaves,[2] five lateral pairs and an anterior unpaired leaf; the anal tube low, truncate, nearly central; the renal pore at its right side. The head rather small; the tentacles digitiform, somewhat flattened. The sides of the body nearly imperceptible; the genital opening contracted.[3] The foot rather strong, somewhat pointed at the end; the anterior margin with a deep furrow, the superior lip rather strong and prominent, cleft in the median line.

The peritoneum with very fine dark points (brown-black) spread everywhere; entirely without true spicules.

The central nervous system showed the cerebro-visceral ganglia somewhat elongate, thicker and broader in the posterior part, nearly not excavated in the exterior margin; the pedal ones of oval form, larger than the visceral. The olfactory ganglia very short-stalked, bulbiform, a little smaller than the buccal; a small optic ganglion, the optic nerve short. At the inferior side of the posterior part of the right visceral (fig 1a) ganglion is a short-stalked (fig. 1b) ganglion genitale giving off several nerves, one of them has at its root another ganglion (fig. 1c). The common commissure not longer than the

[1] Alder and Hancock mention merely ten to fifteen leaves.
[2] Alder and Hancock mention fifteen leaves.
[3] The representation of the penis (?) (l. c. Pl. 3, f. 3) by Alder and Hancock cannot be correct.

transverse diameter of the pedal ganglion, rather strong. The buccal ganglia of roundish form, connected through a very short commissure; the gastro-œsophageal ganglia short-stalked, reaching scarcely one-quarter of the size of the former, with one very large and some smaller cells.[1]

The eyes with black pigment and shining, horn-yellow lens. The otocysts at the slight emargination at the outer margin of the cerebro-visceral ganglia, crammed with otokonia of the usual kind. The broad leaves of the rhinophoria stiffened in the usual way by long, much calcified spicula, perpendicular on the free margin of the leaves. The skin of the back crowded with spicula,[2] mostly very large and much calcified; in the rather low (height 0.5 mm.) granules (fig. 2) crowded erect spicules. In the insterstitial tissue of the intestines true spicula are neither many nor large.

The mouth-tube about 2.0 mm. long, strong, rather wide, quite as usual. The bulbus pharyngeus 3.0 mm. long, with a height of 2.3 and reaching a breadth of 2.5 mm.; the rasp-sheath also projecting 1.0 mm. from the hindermost part of the under side of the bulbus. The form of the bulbus and its retractors as usual; the lip-disk whitish, clothed with a yellowish cuticula. The tongue of usual form; on the shining horny-yellow radula eleven rows of teeth, further backwards twelve developed and four younger rows; the total number of rows thus twenty-seven.[3] The teeth of yellowish color; the height of the outermost 0.06, of the next 0.08 mm.; the height reaches at most about 0.22 mm. The two foremost rows were rather incomplete, in the fourth row were twenty-four, and the number of teeth then increases to twenty-seven.[4] The rhachis (fig. 3a) rather broad. The plates of the usual form,[5] with the usual wing-like expansion of the exterior part of the body and of the root of the hook (figs. 4, 5); the first (fig. 3) with lower hook, which on the succeeding teeth slowly

[1] This representation of the central nervous system in most points agrees with that of Hancock and Embleton (l. c. p. 233, Pl. XVII, fig. 2. 3).

[2] Collingwood (Annals and Mag. of N. Hist., 3 Ser., III, 1859, p. 462) mentions the spicules of this species (from the estuary of the Mersey) as "very elegant, consisting of a broad embossed plate with a double and beautifully serrated edge, terminating abruptly in a blunt apex."

[3] Alder and Hancock mention twenty-four rows, whereof eleven were on the tongue.

[4] Alder and Hancock mention twenty-five plates in the rows.

[5] Cf. my Malacolog. Unters. (Semper, Philipp. II, ii), Heft XIV., 1878, (Asteronotus), p. 636.

increases in height; then the teeth keep the same height and decrease
again in the outer part of the rows (fig. 5); the four to six interior
teeth are more erect, with shorter body and thinner hook (figs. 5, 6).

The salivary glands long, thin, whitish.[1] The œsophagus about
6 mm. long, rather wide, with strong longitudinal folds.[2] The stomach
small, included in the liver; the biliary apertures as usual.

The intestine issues through the liver behind the region of junction of the first and second third of the liver; the first anteriorly proceeding part lodged in a groove on the superior side of the liver, not passing beyond the anterior margin of that organ, about 2.5 mm. in length; the rest of the intestine about 10.0 mm. in length; the diameter of the intestine 0.8–1.3 mm.; the longitudinal folds rather strong.

The liver of yellowish color, more grayish on the surface; 9.0 mm. in length, by a breadth of 5.5 and a height of 4.0 mm.; the posterior end rounded; more than the anterior half of the under side, especially its right part, is excavated (for the anterior genital mass) and behind this is a deep transverse groove. The vesica fellea lying at the left side of the offshoot of the intestine, rather small, in height about 1.25 mm., reaching nearly to the surface of the liver, nearly cylindrical.

The heart as usual. The sanguineous glands whitish, rather flattened; the anterior obliquely triangular with the point, as usual, adhering to the under side of the junction of the two cerebral ganglia; in length 2.0 by a breadth of 1.5 mm.; the posterior transversely elongate-oval, with a breadth of 3.5 by a length of 1.5 mm. The renal syrinx melon-shaped, its largest diameter about 0.75 mm.; its free duct nearly three times as long; a strong continuation of it passing along the floor of the rather large renal chamber, to the region of the pylorus.

The hermaphroditic gland spread in large groups of ramifications over nearly the whole liver and by its brighter yellowish color somewhat contrasted with it; in its lobules were masses of zoösperms and rather small oögene cells. The anterior genital mass[3] in length 5.0 by a breadth of 2.5 and a height of 4.0 mm.; the right side rather convex, meeting the more flattened left side at the sharp superior margin,

---

[1] They are in this way also mentioned by H. and E. (l. c., p. 215, Pl. XII, fig. 2cc).

[2] The dilatation on the œsophagus mentioned and figured by H. and E. (l. c., p. 215, Pl. XII, fig. 2d) could not be seen in the specimen examined by me.

[3] Cf. the Pl. XIV, f. 9, of Hancock and Embleton.

the under side flattened. The ampulla of the hermaphroditic gland resting on the superior posterior part of the genital mass, whitish, making a large curve, about 5.0 mm. long, with a diameter of nearly 1.5 mm. The spermatoduct in its first part, as near as could be determined, rather thick than thin, not very long, forming (fig. 11e, 7e) a little coil on the upper end of the penis. The penis (fig. 7f) cylindrical, curved, about 2.5 mm. long, by a diameter of about 0.8 mm.; the inside with many longitudinal folds; at the upper end of its cavity a low truncated conical prominence (fig. 11b', with a rather wide aperture (fig. 11b), through which opens a little bag (fig. 11), whose inside was clothed with a thin yellowish cuticula, and contained a hollow, nearly colorless tube, that could be extended by tension; it was probably pointed (the point seemed broken off); its length was about 0.9 mm.; the spermatoduct opened (fig. 11a) in the upper part of this bag. Hancock has (l. c. Pl. XIV, fig. 9c, 10; Pl. XV, fig. 1, 2) seen the penis and the "stiletto," but he too seems (l. c. p. 220) not at all clear about these organs. At the side of the opening for the penis in the vestibulum genitale was another aperture which led into a bag, from whose bottom projected a hard, whitish, somewhat compressed conical spur (fig. 7d, 10, that under the influence of nitric acid grew more pellucid, but developed very little gas; through the axis of the organ down to the fine aperture on the point, passes a slender tube (fig. 10), the continuation of the fine coiled duct of the gland of the organ.[1] This gland (glandula hastatoria, fig. 7e, 8d) overlies the upper part of the vagina (fig. 7a, b); it is heart-shaped, of a transverse diameter (breadth) of 2.0, and a thickness of 1.0 mm.; the gland did not contain any larger cavity. The spermatotheca (fig. 8a) whitish, nearly spherical, having a largest diameter of 2.5 mm.; filled with fatty cells and detritus; the two ducts (fig. 8c, e) as usual, the vagina rather wide (fig. 7a, b), with longitudinal folds on the inside. The spermatocysta yellowish, spherical, 1.5 mm. in diameter (fig. 8b), filled with zoösperms; short-stalked. The mucous gland not forming quite half of the anterior genital mass, consisting of a smaller anterior biconvex part, and a large flattened wing-like posterior part; the space between them nearly filled by the spermatotheca

---

[1] These organs, the gland and the spur, have also been seen (l. c., Pl. XV, fig. 9) by Hancock, but he does not mention them (in the text, and explanation of the figures). In another of his figures (fig. 10b) the spur is designated (l. c., p. 248) as "male intromittent organ," and the (fig. 10e, f) true penis as "penis-like organ furnished with a stiletto."

and the spermatocysta, the color of the gland yellowish-white, on the left side of the anterior part a central yellow mass; the duct of the mucous gland rather short.

All the former genera of *Dorididæ* belonged to the large group of *Dorididæ cryptobranchiatæ*;[1] the following are to be registered in the group of *Dorididæ eleutherobranchiatæ* (*D. phanerobranchiata*). This section is also characterized by the non-retractility of the gill, by a sucking-crop connected with the bulbus pharyngeus and by a peculiar armature of the tongue, consisting usually of a single large lateral plate and a single or several outer plates. This group seems chiefly limited to northern climes, and contains at present the genera *Akiodoris*, *Acanthodoris*, *Adalaria*, *Lamellidoris*, *Goniodoris* and *Doridunculus*,[2] also *Ancula*, *Drepania*[3] and *Idalia*.

## AKIODORIS, Bergh.

*Akiodoris*, Bgh. Gattungen nordischer Doriden, l. c., 1879, p. 354.

Forma ut in *Lamellidoridibus*. Nothæum supra granulosum. Branchia non retractilis, e foliis tripinnatis non multis et ad modum ferri equini positis formata. Caput latum, veliforme; tentaculis brevibus, lobiformibus. Aperturæ rhinophoriales integræ.

Discus labialis sine armatura. Ingluvies buccalis bulbo connata. Radula rhachide quasi nuda; pleuris dentibus lateralibus depressis non multis; (12-13) quorum duo intimi fortiores, quasi subhamati. Penis glande uncis simplicibus, furcatis vel palmatis armatus. Vagina indumento valloso peculiari instructa.

The animals belonging to this group resemble externally especially the *Lamellidorides*. The back is finely granulated; the head large, veil-shaped, with short tentacles, which are lobate and pointed. The openings of the rhinophor-holes with plain margins, surrounded by several larger papillæ. The non retractile branchia nearly horseshoe-shaped, consisting of a mediocre number of leaves. The lip-disk

---

[1] Cf. my "Gattungen nordischer Doriden," l. c. p. 341.

[2] The genus *Doridunculus* of G. O. Sars (Moll. regionis arcticæ Norveg., 1878, p. 309. Tab. 27, fig. 2a-d, Tab. XIV, fig. 5), which externally approaches *Goniodoris* and other *Dorididæ eleutherobranchiatæ* in the character of the radula, is hitherto only known from the northeastern part of the Atlantic (Lofoten), and by a single species (*D. echinulatus*, S.).

[3] In the *Ancula* and *Drepania* the penis is armed as in so many *Doridida* with a series of small hooks.

without armature. The tongue with transverse thickenings of the rhachis; the lateral plates somewhat depressed; the two first different from the rest, larger and with a denticle at the root of the hook; the rest without any such, the external quite without a hook. A sucking-crop on the upper side of the bulbus pharyngeus, but sessile, depressed conical, and not consisting of two symmetrical halves. The large stomach free on the surface of the liver. The glans of the long penis with a strong and quite peculiar armature, consisting of strong hooks, partly simple, partly bifurcate and partly digitate, with strong digitations. The vagina with a peculiar armature of high palisades.

This interesting genus externally most resembles the *Lamellidorides*, both in reference to the nature of the back, to the form and size of the gill and in the want of armature of the lip-disk; the region of the openings of the rhinophor-holes differ in the want of a glabella and by the presence of a larger number of surrounding papillæ. The genital opening somewhat recalls the *Acanthodorides*, as do also the (tripinnate) branchial leaves and the sucking-crop, but this is not divided in two distinct halves as in this last genus. The armature of the tongue is very different from that of the *Lamellidorides*. *Adalariæ* and *Acanthodorides*; the large hook-formed lateral plates of these genera are wanting, and in their places are two large depressed lateral plates, with small hooks; the external plates somewhat recalling those of the *Adalariæ*; the rhachis rather broad, with transverse thickenings of the cuticula, corresponding to the rows of plates. In the very peculiar form of armature of the glans penis, and by the peculiar clothing of the vagina, the *Akiodorides* differ from all the above-cited genera.

Only a single species of the genus is hitherto known, the new one, that will be described below.

1. *Ak. lutescens*, Bgh., n. sp. Oceanum Pacificum.

1. **Ak. lutescens**, Bgh., n. sp. Pl. IV, fig. 3; pl. V, fig. 11-14; pl. VI, fig. 1-20; pl. VII, fig. 1-8; pl. VIII, fig. 1-2.

Color lutescens.
*Habitat.* Oceanum Pacificum septentrion. (Nazan Bay).

Of this form I have had a large single specimen for examination, obtained in August, 1873, by Dall, on stony bottom, at low water, in Nazan Bay, Atka Island, Aleutians.

According to Dall, the color of the living animal was "yellowish-white;" preserved in spirits, it was of a uniform dirty yellowish color.

The length was 32.0 mm., by a breadth of 19.0 mm., and a height of 13.0 mm.; the breadth of the foot 12.5 mm., of the mantle-brim 3.0 mm.; the height of the rhinophoria 3.0 mm., of the branchial leaves 2.5 mm.; the length of the genital opening 2.25 mm.

The form was elongate-oval, somewhat larger than that of the *Lam. bilamellata*. The papillæ of the back relatively smaller and more rounded than in that animal. The openings of the rhinophor-holes an oblique oval slit; the margins plain; several (six to eight) larger papillæ (of about 1.0 mm. in height) in the immediate vicinity of the holes; the club of the rhinophoria with about thirty leaves. The branchia with about ten leaves. The anal papilla low, with a stellate aperture; the renal orifice as usual; the interbranchial space crowded with rather pointed and high papillæ. The head and tentacles as in allied forms. The genital papilla of oval form, with a large, longitudinal, crescentic slit. The rather broad foot with the usual anterior marginal furrow. The peritoneum colorless, without spicula.

The central nervous system more flattened than in allied forms; the cerebro-visceral ganglia reniform, a little broader in the anterior part; the pedal ganglia less flattened than the former, larger than the visceral ones, of oval form, on the outside of the cerebro-visceral. The proximal olfactory ganglia a little smaller than the buccal ones, bulbiform; distal ganglia could not be found. The commissure not broad, not short. The buccal ganglia of oval form, closely connected; the gastro-œsophageal roundish, rather long-stalked, in size about one-sixth of the former, with one large cell and several (three or four) smaller ones.

The nervi optici rather long; the eyes with yellowish lens and black pigment. The otocysts in the usual place, filled with otokonia of the usual kind. The leaves of the club of the rhinophoria very richly furnished with thick (diameter. 0.04 mm.) and long spicula, more or less calcareous, and very often giving off a thick twig of greater or less length (Pl. V, fig. 12); for the most part set perpendicularly or obliquely on the free margin of the leaves. The axes of the organs and the short stalk stuffed with strong and very much calcified spicules. In the skin of the back a mass of spicula of the same kind (Pl. IV, fig. 13) as above, or still more hardened; the papillæ of the back solidified in the usual way (Pl. V, fig. 11). In the interstitial tissue fewer and smaller spicules.

The oral tube rather short, wide. The bulbus pharyngeus of usual form, about 5.5 mm. long by a height of 4.5 mm., (and at the upper part of the sucking-crop of 5.5 mm.), and a breadth of 4.75 mm.; the sheath of the radula projecting about 1.3 mm. backwards and downwards. The lip-disk large, clothed with a thick yellow cuticula; the true mouth forming a narrow vertical slit. The cap-shaped sucking-crop almost exactly as in *Ac. pilosa*, but more conical and without external signs of duplication; on the inside clothed with a yellowish cuticula, opening into the buccal cavity through a wide slit. The tongue rather broad; on the fine reddish-yellow colored radula seventeen rows of teeth, also on the point of the tongue were traces of six entirely vanished rows; the two first rows very incomplete, reduced to some external plates. Further backwards were seen forty-two developed and three younger rows, or, all in all, the animal presented sixty-two rows of teeth. The most external plate of each row is quite colorless, the next two or three pale yellowish, the following all of horny-yellow color; the rhachis colorless. The length of the most external plate about 0.035 mm., of the next about 0.05 mm., of the following 0.07 mm.; the length of the second large plate about 0.2 mm., of the first 0.022 mm.; the breadth of the rhachis about 0.22 mm. The rhachis thickened between the rows and forming arched elevations between them (Pl. VI, fig. 1*a*, 3; Pl. VIII, fig. 1*a*). The first two plates rather large (Pl. VI, fig. 1*bb*, *cc*, 4–6; Pl. VIII, fig. 1*b*, *c*); with a short strong hook and a stout denticle at each side of it, the outer denticle broader; the hook of the second plate somewhat larger than that of the first; sometimes a slight crenulation on the outer margin of the first plate (fig. 5). All the following ten or eleven plates (Pl. VI, fig. 2*e*, *f*; Pl. VIII, fig. 2*a*, *b*) of the same type, by degrees decreasing in size, consisting of a quadrilateral basal part, from which (Pl. VI, fig. 7–13), in most of them, rises a strong, short, broad hook; the two or three outmost plates (Pl. VI, fig. 2*f*; Pl. VIII, fig. 2) formed of the basal part alone; the rest with the hook gradually more developed.

The salivary glands yellowish-white, flattened, ribbon-shaped, of about 10.5 mm. in length, reaching to the cardia, where they are agglutinated one to another; the breadth in the foremost part about 0.75 mm, in the middle 1.5 mm., the posterior part again somewhat narrower; the duct of the gland rather short.

The œsophagus rather wide, about 13.0 mm. long, the inside with rather strong longitudinal folds; it opens into the stomach, which lies free in a cleft on the upper side of the liver. This organ (Pl. VI, fig. 17*a*) is of oval form, of about 6.5 mm. largest diameter; the inside with rather strong longitudinal folds; the pylorus (fig. 17) in the neighborhood of the cardia. The intestine advancing from the stomach to the fore-end (fig. 17*b*) of the liver, in this part about 10.0 mm. long; forming a knee and retrograding to the anal nipple in a length of 23.0 mm. The contents of the stomach were indeterminable animal matter, mixed with some diatomaceæ.

The liver 20.0 mm. long by a height of 10.0 mm. and a breadth of about 12.0 mm.; the posterior end rounded; a little more than the anterior half of the under side obliquely flattened (by the anterior genital mass) showing the cardiac end of the œsophagus and the root of the hermaphroditic duct. On the anterior part of the upper surface is a cleft for the stomach and for the biliary sac; the color of the surface and of the substance of the liver is grayish-yellow. The biliary sac (fig. 17*c*) lying before the stomach, on the right side of the intestine, large as the stomach), somewhat flattened, grayish, of rounded outline and about 4.5 mm. largest diameter; the contents, as in the stomach.

The heart as usual. The sanguineous gland whitish, entirely covering the nervous system, about 6.0 mm. long, by a breadth of 4.5 and a height of only 1.0 mm.

The hermaphroditic gland yolk-yellow, covering the upper side of the liver with a thick layer; in its lobes large oögene cells and masses of zoösperms. The anterior genital mass large, about 14.0 mm. long by a breadth of 9.0 and a height of 11.0 mm., flattened and a little excavated on the left side, with an excavation on the fore side, the right side very convex. The hermaphroditic duct whitish, rather thin (diameter about 0.75–1.0 mm.), passing straight over the left side of the genital mass to its anterior end, without formation of any (distinct) ampulla. The first part of the spermatoduct whitish, forming several long windings on the upper part of the forepart of the mass and passing into the yellowish (Pl. VI. fig. 18*a*) continuation; this, with its numerous coils, forms a large flattened layer on the fore-end of the right side of the mass; it then rather suddenly passes into a much thinner whitish continuation (fig. 18*b*) about 6 mm. long, that slopes (fig. 18*c*) into the penis, which (retracted) was lying on the lowest anterior part of the right side of the mass. The penis was cylindrical, of the length of 11.0 mm. by a diameter of 1.5 mm.; the truncated, cylindrical, yellowish (under a magnifier nodulous) glans forming (Pl. V, fig. 13, 14) a prominence of the length of nearly 1.0 mm. in the vestibulum. This glans was partly covered on the outer side (fig. 13, 14), but especially on the margin of the wide, gaping orifice and on its inside for a length of about 4.0 mm. (Pl. VII, figs. 2–4). with rather crowded and apparently irregularly set claws. The claws were very strong and for the most part broad and high (fig. 3, 4), even reaching a height of about 0.3 mm. (fig. 4). In the interior of the glans, especially in its posterior part (fig. 5c), the claws were less broad and simply uncinate or bifurcated, otherwise mostly broader and with digitations of the margin. The body of the claws was plain or curved; the end simply pointed, bi- or trifurcate or with digitations, sometimes very strangely formed. They consisted of a cuticula and its matrix; very often, especially on the outside of the glans, the cuticula was torn off and the (fig. 20) rounded or pointed naked matrix was left. The whitish spherical spermatotheca (Pl. VI, fig. 19a) was about 3.5 mm. in diameter, laterally communicating through a short petiolus adhering to the upper end of the vagina, with a sinuosity into which opens the elongate, yellowish spermatocysta (fig. 19b), which had a length of about 2.0 mm., and from which issues the long duct of the mucous gland (fig. 19c). The grayish vagina very strong (fig. 18e). about 7.0 mm. long, elongate-conical; the lowest part wide, having a diameter of about 3.25 mm.; the walls thick, with a very peculiar internal lining, consisting of cylindrical palisades (Pl. VII, fig. 6–8) of a height of about 0.4 by a greatest diameter of 0.07–0.08 mm.; between the larger were seen smaller and very small ones. The palisades seemed to be densely clothed (fig. 8) with cilia, and showed a nearly colorless axis (fig. 6, 8) up to their points; the axes were often denuded (fig. 6) after the sheath has been torn away. This lining continued up to the superior end of the vagina, but not beyond it.

The mucous gland large, whitish, and yellowish-white; the anterior half yolk-yellow, denuded on the fore-end of the genital mass; the duct short.

A variety (Pl. VI, fig. 14–20) of this species has also been found by Dall, in July, 1873, at low water, in Kyska Harbor (Aleutians). According to Dall the color of the living animal was "yellowish." The animal preserved in spirits was of a uniform light yellowish color. The length about 18.0 mm. by a breadth reaching 8.0 mm. and a height of 6.0 mm.; the breadth of the foot at the fore-end 5.0 mm., the margin of the mantle freely projecting 1.5 mm.; the height of the rhinophoria 1.5 mm., of the branchial leaves 1.5 mm. Around the plain margins of the rhinophor-holes seven to nine large conical tubercles; the club of the rhinophoria with about twenty leaves. Around the branchial ring, as well as in the centre of it around the vent, rather large conical tubercles 1.5 mm. in height; the branchial leaves, fifteen in number, as far as could be determined.

The oral tube strong, 4.5 mm. long, wide. The bulbus pharyngeus about 5.5 mm. long, by a height of 3.0, and a breadth of 3.75 mm.; the rasp-sheath about 1.75 mm., freely projecting, bent upwards. The cuticula of the lip-disk yellowish. The tongue with about thirty-five rows of plates (fig. 14–16); further backwards, twenty-five developed and four younger rows; the total number of rows sixty-four. On the posterior part of the tongue fourteen plates, the number increasing backwards to fifteen or sixteen. The five anterior rows very incomplete, only represented by 1. 7. 9, 10, 12 plates (on each side). The plates as above. The breadth of the rhachis reaching to about 0.17 mm. The glandulæ salivales 6.0 mm. long. The stomach (fig. 17a) about 4.0 mm. long. The contents of the digestive cavity a mass of sponge. The vesica fellea (fig. 17c) about 2.5 mm. high, with strong folds on the inside. The anterior genital mass quite as above, also the spermatotheca and the spermatocysta (fig. 19), the penis (fig. 18, 20), and the vagina (fig. 18, 19).

## LAMELLIDORIS, Alder et Hancock.

*Lamellidoris*, A. et H., Monogr. Brit. Nudibr. Moll., Part VII. 1855, p. xvii.
*Lamellidoris*, A. et H., R. Bergh, Malacolog. Untersuch. (Semper, Philipp. II, ii), Heft xiv, 1878, p. 603-615.
*Lamellidoris*. A. et H., R. Bergh, Gatt. nörd. Doriden, l. c., 1879, p. 362-365.

Corpus vix depressum, nothæo granulato. Branchia (non retractilis) e foliis (multis) simpliciter pinnatis, ut plurimum in formam ferri equini dispositis, formata. Caput latum, semilunare, angulis tentacularibus. Aperturæ rhinophoriales, margine integro; tuberculis anticis 2–3, calvitie postica.

Cuticula aperturæ oralis infra asserculis duobus incrassata, et ante annulus papillarum angustus. Lingua rhachide lamellis humilibus instructa; pleuris dente interno hamiformi permagno et externo compresso lamelliformi unco minuto prædito armatis. Ingluvies buccalis (suctoria) petiolo bulbo pharyngeo connata, tympaniformis.

Penis apice (glande) curvatus, non armatus. Vagina brevis.

The genus *Lamellidoris* was established (1855) by Alder and Hancock, to receive two small groups of *Doridida*, one with the *D. bilamellata* as type, to which especially the name of the group is here restricted; and the other, characterized by a more depressed form and the naked rhachis of the tongue, with the *D. depressa*, A. et H., as type. Hancock has given some anatomical remarks on the typical form (*D. bilamellata*, L.); but nothing else had been since made known about these animals[1] until my just cited notice and those of G. O. Sars.[2]

The *Lamellidorides* approach the *Acanthodorides*, but differ even here, externally, by the coarsely granulated surface of the back and by the larger number of the branchial leaves, which are set in the form of a horseshoe; the openings of the rhinophor-holes, the tentacles as well as the genital opening are also of a different shape. More notable still are the anatomical differences; the *Lamellidorides* want the armature of the lip-disk, which is found in the other group; the armature of the tongue is quite different (1, I—1—I, 1), and the buccal crop is connected with the bulbus pharyngeus by a stalk. The penis is quite different from that of the *Acanthodorides*, and without true armature; the vagina is short. After all the *Lamellidorides* are much more allied to the *Adalariæ*.

The form of the body, as in the *Acanthodorides*, not very depressed. The back covered all over with semi-globular and short club-formed papillæ. The openings of the rhinophor-holes with plain margins and

---

[1] According to H. & A. Adams (the Gen. of Recent Moll., II, 1858, p. 657), *Lamellidoris* is a synonym of "*Onchidoris*, Blv.," which name is employed by Adams for a group, whose type should be *D. pusilla*, A. et H. (that scarcely belongs to the true *Lamellidorides*). Cf. also Gray, Guide I, 1857, p. 207.

The genus *Onchidoris* of Blainville (Man. de Malac., 1825, p. 489, Pl. XLVI, f. 8.), ought to be rejected entirely, as founded very likely only on bad observation; the genus figures with nearly impossible characters, both in relation to the tentacles ("quatre tentacules comme dans les *Doris*, outre deux appendices labiaux") and to the anus ("médian à la partie inférieure et postérieure du rebord du manteau"). The type of the genus Blainville found in the British Mus. (London), where it seemed to have disappeared, at least it was not to be found in the collection of nudibranchiates which I looked over in May, 1873; while, on the contrary, I found the long-lost type of the genus *Linguella*, Blv., in his original glass, and so have re-established the denomination *Linguella* for the much later (1861) *Sancara*, Bgh. Cf. my Malacolog. Unters., Heft vi, 1874, p. 248). Later, Mr. Abraham (l. c. p. 223) seems to have found the original specimen again.

[2] G. O. Sars, Moll. reg. arct. Norv., 1878, p. 306. Tab. XIII, figs. 5, 6; Tab. XIV, fig. 2, 3.

commonly two larger papillæ before and a bare space behind them. The gill (not retractile) consisting chiefly of several (usually 20–30) tripinnate leaves, set in the form of a horseshoe. The head large, veil-formed (semilunar), with produced and pointed side-parts, which are adherent to the foot nearly to the point. The genital openings not being a slit, but on a large tubercle.

The cuticula of the oral aperture is thickened below, near the median line, into a ledge; and on the outside is a ring of hard papillæ. The buccal crop, connected through a petiolus with the foremost part of the upper side of the bulbus pharyngeus, is drum-shaped; on the inside clothed with a strong cuticula. The tongue has on the rhachis short compressed lamellæ, on each side of these is a very large upright plate with large compressed body and a hook which on the inside is either plain or denticulated; at the outside of this plate is another, compressed but much smaller and with a little rudimentary hook. The salivary glands forming a short, coiled mass at each side of the root of the œsophagus. The œsophagus without diverticle at its origin. The spermatoduct (as in the *Acanthodorides*) very long; the penis short, its glans curved and clothed with a rather thick cuticula, but otherwise not armed. The spermatocysta imbedded in the mucous gland;[1] the vagina short.

About the biological relations of the animals belonging to this group very little is hitherto known. Where the species occur, they seem to be rather abundant in individuals (cf. about the *Lam. bilamellata*, Collingwood, in Ann. Mag. Nat. Hist., 3 S. III, 1859, p. 163). The spawn of several species (*L. bilamellata, L. diaphana, L. inconspicua, L. aspera, L. depressa, L. pusilla*) has been described by Alder and Hancock, and that of a single species (*L. muricata*) by Sars, Meyer and Moebius, etc. The first stages of the development of this last form have been followed by Sars.[2]

The group seems limited to the northern part of the Atlantic and of the Pacific. To the same belong with certainty some properly examined species, and, besides, several others mentioned in the literature can, with more or less probability, be referred to it.

---

[1] The spermatocysta has not been seen by Alder and Hancock. Cf. l. c., 1852. Pl. XIV, fig. 8 (p. 219).

[2] Archiv. für Naturges, 1840, p. 210, Tab. 7.

## A.

1. *L. bilamellata* (L.).   Oc. Atlant.
2. *L. varians*, Bgh., n. sp.   Oc. Pacif.
3. *L. hystricina*, Bgh., n. sp.   Oc. Pacif.
4. *L. muricata* (O. Fr. Müller).   Oc. Pacif.
5. *L. diaphana* (Ald. et Hanc.).   Oc. Atlant.
   *D. diaphana*, A. et H., Monogr. Part ii, fam. 1, Pl. 10; Part vii, Pl. 46 suppl. fig. 9.
6. *L. aspera* (A. et H.).[1]   Oc. Atlant.
   *D. aspera*, A. et H., l. c., Part v, fam. 1, Pl. 2, fig. 15; Part vi, fam. 1, Pl. 9, fig. 1-9; Part vii, Pl. 46, suppl. text; Pl. 48, suppl. fig. 2.

## B.

7. *L. sparsa* (A. et H.).   Oc. Atlant.
   *D. sparsa*, A. et H., l. c., Part iv, fam. 1, Pl. 14; Part vii, Pl. 46, suppl. text.
8. *L. depressa* (A. et H.).   Oc. Atlant.
   *D. depressa*, A. et H., l. c., Part v, fam. 1, Pl. 12, fig. 1-8; Part vii, Pl. 46, suppl. fig 12.
   ? *Villiersia scutigera*, d'Orb., Mag. de Zool., 1837, p. 15, Pl. 109, fig. 1-4.
9. *L. inconspicua* (A. et H.).   Oc. Atlant.
   *D. inconspicua*, A. et H., l. c., Part v, fam. 1 Pl. 12, fig. 9-16; Part vii, Pl. 46, suppl. fig. 13.
10. *L. oblonga* (A. et H.).   Oc. Atlant.
    *D. oblonga*, A. et H., l. c., Part v, fam. 1, Pl. 16, fig. 4 5; Part vii. Pl. 46, suppl. fig. 10.
11. *L. pusilla* (A. et H.).   Oc. Atlant.
    *D. pusilla*, A. et H., l. c., Part ii, fam. 1, Pl. 13; Part vii, Pl. 46, suppl. text; app. p. iii.
12. *L. luteocincta* (M. Sars).[2]   Oc. Atlant.
13. *L. (?) ulidiana* (Thomps.).   Oc. Atlant.
    *D. ulidiana*, Th., Ann. Mag., Nat. Hist., xv, 18, p. 81.
    *D. ulidiana*, Th., Ald. et Hanc., l. c., Part vii, p. 42, app. p. ii.
14. *L. (?) tenella* (Agassiz).   Oc. Atlant.
    *D. tenella*, Ag., Gould, Rep. on the Inv. of Massachusetts, ed. Binney, 1870, p. 229, Pl. xx, fig. 289, 290, 293.
15. *L. (?) pallida* (Ag.).   Oc. Atlant.
    *D. pallida*, Ag., Gould, l. c., p. 229, Pl. xx, fig. 284, 287, 288, 291.

---

[1] According to Mörch (Synopsis Moll. mar. Daniæ, Vidensk. Meddel. fra naturh. Foren. i Kbhvn., 1871, p. 179) this species ought to be identical with the *D. muricata* of Meyer and Moebius; but this is, of course, impossible.

[2] The organs of the bulbus pharyngeus of this species have just been figured by G. O. Sars (Moll. reg. arct. Norv., 1878, Tab. xiv, fig. 3).

16. *L. (?) diademata* (Ag.). Oc. Atlant.
    *D. diademata*, Ag., Gould, l. c., p. 230, Pl. xxi, fig. 298, 300, 301-304.
17. *L. (?) grisea* (Stimps.). Oc. Atlant. Gould, l. c., p. 232, Pl. xx, fig. 292, 295.
18. *L. (?) derelicta* (Fischer). Oc. Atlant.
    *D. derelicta*, F., Journ. de conchyl., xv, 1867, p. 7.
19. *L. (?) tuberculata* (Hutton). Oc. Pacif. (Nova Zeland.).
    *Onchidoris tuberculatus*, Hutton, cf. Abraham, l. c., p. 226.
20. *L. (?) eubalia* (Fischer). Oc. Atlant.
    *Doris eubalia*, F., Journ. de conchyl., xx, 1872, p. 10.

1. **L. bilamellata** (L.), var. *pacifica*. Plate V, fig. 10; Plate XI, fig. 3-9.

Color albido-flavescens, maculis fuscis plus minusve variegatus.
Dentes laterales margine laevi.
*Hab.* Oc. Pacific. sepentr. (Mar. Beringi).

Six specimens of this variety of the Atlantic species were taken by Dall, in Bering Sea (Hagmeister Id.), in August, 1874, at low water, on a gravel beach. Three were sacrificed for the anatomical examination.

According to Dall, the color of the living animal was "yellowish-white with brown maculae."

The length of the specimens preserved in spirits was 11-13.0 mm. by a height of 4.5-5.5 mm. and a breadth of 6-10.0 mm.; the height of the rhinophoria 1.75-2.2, of the branchial leaves 1-1.2 mm.; the breadth of the foot at the fore-end about 5-8.0 mm.; the margin of the mantle projecting freely about 1.5-2.0 mm. The color of the individuals on the back was yellow-white, marmorated with light reddish-brown, this marbling always occupying the spaces between the tubercles, which are nearly white (or light yellowish); the branchial leaves of the same reddish color; the club of the rhinophoria yellowish-white; the under side of the body yellowish-white or whitish.

The form was elongate-oval. The head flattened, nearly semicircular, with the tentacular edges a little prominent. The vicinity of the posterior margin of the rhinophor-holes plain, at the anterior two large erect tubercles; the club of the rhinophoria with about twenty leaves, the stem rather short. The back covered all over with semi-globular and short club-shaped rounded tubercles of different sizes, mostly small, mixed with many larger ones 0.75 mm. in diameter; the larger tubercles mostly showing a spinous surface (Pl. V, fig. 10)[1] when magnified.

[1] Cf. my "Malacolog. Unters." (Semper, II, ii) Tab. LXVIII, fig. 15-16.

The openings of the rhinophor-holes and of the branchial area (fig. 3*bb*) surrounded by large and small tubercles which also were spread over the central part of it (fig. 3). The branchial leaves (fig. 3*aa*) were about twenty-four or twenty-five in number, set in a transverse reniform ring; the leaves in the front part much larger than the rest. The anus as usual, scarcely projecting. The under side of the margin of the mantle quite smooth. The genital openings always quite contracted. The foot large, with a fine line along its anterior margin.

The cerebro-visceral ganglia short-reniform; the pedal ones not much smaller, of oval form, set nearly at a right angle to the inferior face of the former; the olfactory ganglia bulbiform or ovoid. The buccal ganglia rather flattened, of roundish contour, a little larger than the olfactory ones; the commissure between them very short; the gastro-œsophageal ganglia not very short-stalked, roundish, in size about one-quarter of the buccal ganglia, with three large cells. The three commissures very distinct, the sub-cerebral and the pedal connected throughout most of their length; the visceral thin, not giving off a genital nerve.

The eyes with black pigment, yellowish lens; the nervus opticus nearly as long as half the breadth of the cerebral ganglion. The otocysts as large as the eyes, crowded with otokonia of the usual kind. The leaves of the rhinophoria without spicules; the axis of these organs, on the other hand, were filled with such spicules, partly circularly and concentrically arranged. The tubercles of the back stuffed with ordinary spicules (fig. 10) in the usual way, the larger spicules mostly very prominent on the surface.

The oral tube as usual. The bulbus pharyngeus of the usual form, about 2.0 mm. long; the lip-disk with a rather thick yellowish cuticula, and inwards with the same belt of (about ten to fifteen) rows of small denticles as in the *L. hystricina* (cf. below); the sheath of the radula somewhat bent upwards, freely projecting behind the bulbus for as great a length as that of the bulbus itself. The tongue (in the three individuals) with ten or eleven series of plates, in the sheath ten or eleven developed and three younger rows; the total number of rows being thus twenty four or twenty-five. The plates light yellowish in their thicker parts, otherwise nearly colorless. The length of the median plates reaching about 0.12 mm., the height of the external ones 0.10 mm. The median (fig. 7*a*) and exterior plates (fig. 7*b*) quite as usual; the large ones of the usual forms (fig. 7*b*'), sometimes, especially the foremost, with rather obtuse point (fig. 9). The buccal crop (fig. 4, 5) as large as the bulbus, of quite the usual form, rather petiolate.[1]

The salivary glands forming (on each side) a large, thick, whitish mass between the bulbus and the central nervous system (with the glandulæ sanguineæ).

The œsophagus rather wide. The stomach and the intestine as usual. The liver as usual, much flattened on the right anterior half.

The heart rather large. The gland. sanguineæ large, whitish, covering the upper side of the central nervous system, the foremost part in one individual very narrow. The renal syrinx about 1.0 mm. long, with strong longitudinal folds, its clothing as usual.

The anterior genital mass 4–4.5 mm. long by a breadth of 1.25–1.5 and a height of 3–3.3 mm., yellow-white, plano-convex; the anterior, and partly the superior portion formed by the coils of the whitish spermatoduct; in one individual one coil embraced the sheath of the radula. The first part of the spermatoduct strong, when unrolled about 25.0 mm. long; the succeeding part of the length of 4–5.0 mm., thinner; the rest about 7.0 mm. in length, stronger, nearly as in the first part. In the beginning of this last part the true spermatic duct was rolled up in tight coils, the remaining part of its length was nearly straight. The penis about 1.5 mm. long, with the usual glans in the interior. The spermatotheca (fig. 6a) spherical, its chief duct nearly twice as long as the bag, the vagina short (fig. 6c). The spermatocysta appeared pyriform (fig. 6d).

In color this form seems to differ from the typical one, as that is represented by Alder and Hancock (Monogr., Part vi, 1854, fam. 3, Pl. 9); in the anatomical relations no specific differences could be detected.

A specimen of another variety was obtained by Dall, on a gravel beach, at low water, in June, 1874, at Port Etches (Prince William Sound . According to Dall, the mantle was of " brown " color.

The specimen had a length of 13.0 mm., by a breadth of 8.0 mm., and a height of 5.0 mm.; the height of the leaves of the gill was about 1.0 mm. The color of the back was brownish and yellowish; that of the gill, as well as of the rhinophoria, yellowish. The number of leaves of the gill was about thirty.

The bulbus pharyngeus about 1.75 mm. long, by a height of 1.5 mm.; the sheath of the radula nearly as long as the bulbus; the buccal crop

---

[1] In one specimen the form of this organ was entirely as figured in my Malacolog. Untersuch. (Semper, Reisen). Tab. LXV, fig. 2.

a little larger than the bulbus. The radula brownish-yellow, with nine rows of teeth, further back fifteen developed and two younger rows, the total number being twenty-six. The teeth quite as above, dark, horn-colored in their thicker parts; the median ones reaching a height of 0.16 mm. The salivary glands as above-mentioned.

The biliary sac uncommonly small. The black contents of the rectum consisting of undeterminable animal matter, mixed with larger and smaller pieces of small crustacea. The liver much flattened on the right anterior half.

The anterior genital mass large, about 7.0 mm. long, 5.0 mm. high, and 3.0 mm. thick. The ampulla of the hermaphroditic duct whitish, forming a long ansa, about 5.0 mm. long. The spermatoduct shorter than in the other form, otherwise, with the penis, as in that form. The spermatotheca yellowish, short, sac-shaped, of a largest diameter of 3.0 mm.; the spermatocysts about 0.3 mm. long, pyriform. The mucous gland chalk-white and brownish-gray.

Of another variety, Dall, in August, 1872, obtained six specimens, in Sanborn Harbor (Shumagin Ids.), on stony bottom, at low water.

According to Dall, the color of the back of the living animal is "red-brown, with whitish papillæ." The color of the backs of the specimens preserved in spirits was rather uniformly, dirty brown-yellowish, commonly much lighter on the middle, the papillæ whitish; the gill and the rhinophoria of the color of the back; the under side of the whole body yellowish; more whitish on the mantle. The length of the animals varied from 18.0 to 25.0 mm., by a breadth of 11.0 to 16.0 mm., and a height of 8.0 to 12.0 mm.; the breadth of the foot 7.5 to 12.0 mm.; the height of the rhinophoria reaching 3.0 mm., that of the gill 2.0 mm. The form as usual. The horseshoe shape of the gill very pronounced, the number of leaves, twenty-eight to thirty. The gill was surrounded by higher papillæ, which, in the largest specimen, reached the height of about 2.5 mm.; the space inclosed by the gill closely set with similar papillæ, the largest (as large as the above mentioned) in the periphery. The gill can be so deeply drawn back in its groove, that these external and internal papillæ shut over and quite conceal it; the papillæ of the centre smaller; a crest or some few papillæ in the median line go from the anus backwards, between the incurved ends of the gill. The anus small, very slightly prominent; the renal pore on the right side. The openings of the rhinophor-holes as usual, before them the two usual papillæ, behind them a bare space. The papillæ of the back quite as in the previously examined form, the largest (in the largest specimen) reaching the height and the diameter of about 1.5 mm., those in the neighborhood of the gill somewhat larger.

Two smaller individuals were dissected, the larger being harder than these and not so suitable for that purpose. The peritoneum was colorless.

The central nervous system just as in the former specimens, but the buccal ganglia smaller than the olfactory, and the gastro-œsophageal short-stalked.

The eyes as above. The otocysts, under the glass, very distinct as chalk-white points on the hinder and outermost part of the cerebral ganglia. The leaves of the rhinophoria without spicula. The skin and the papillæ of the back as above or still more crowded with very hard spicula.

The oral tube large, (in both individuals) about 2.5 mm. long. The bulbus pharyngeus of the usual form, (in both individuals) about 3.0 long, by a breadth of 1.8 mm, and the height nearly the same; the sheath of the radula projecting straight backwards 2.0 mm. The buccal crop, lying to the left side of the bulbus, somewhat compressed, of about 3.0 mm. largest diameter, the stalk nearly half as long as the largest diameter of the crop. The tongue with ten rows of teeth, further backwards also eleven or twelve developed and three younger rows, the total number thus being twenty-four or twenty-five. They were entirely as in the form first examined.

The salivary glands, the pyloric part of the intestine, with its biliary sac, and the liver as usual. The sanguineous gland whitish, much flattened, covering the whole upper side of the bulbus pharyngeus and the central nervous system; a flattened cavity in its interior. The hermaphroditic gland, through its more reddish color, contrasting with the grayish color of the liver.

The anterior genital mass 11.0 to 12.0 mm. long, by a height reaching 7.0 to 8.0 mm., and a breadth of 4.0 to 4.5 mm. The ampulla of the hermaphroditic duct lying transversely on the lowest and most anterior part of the back of the mucous gland, rather straight or forming nearly a circle, about 5.0 to 7.0 mm. long, whitish. The spermatoduct making many coils on and before the anterior part of the mucous gland; the first part about 35.0 to 45.0 mm. long, the second nearly 25.0 mm. long; the penis about 1.5 to 2.0 mm., projecting freely from the vestibulum, conical; the glans seemed rather short. The spermatotheca of about 3.0 mm. diameter, whitish. The spermatocysta (fig. 6b) quite imbedded in and concealed by the mucous gland, only a part of its chief duct free on the surface of this last; the spermatocysta scarcely shorter than the spermatotheca, pear-shaped, incurved; the duct to the mucous gland (fig. 6d) passing from the end of the bag, the other strong, longer (fig. 6c), opening in the duct of the spermatotheca, where it begins to be wider (vagina); the vagina (fig. 6e) rather wide, but short. The mucous gland whitish, yellowish and dirty yellow.[1]

2. **L. varians,** Bgh. Pl. XI. fig. 13, 14; Pl. XIII, fig. 1.

*L. varians*, B. R. Bergh, Malacol. Unters. l. c., 1878, p. 613, 614.

Color cœrulescens vel albescens vel flavescens.
Dentes laterales margine interno denticulati fere usque ad apicem.
*Hab.* Oc. Pacif. (Ins. Kyska).

Of this species six specimens were taken by Dall, in July, 1873, at Kyska Island, on sandy ground, at a depth of 9–14 fathoms. Four specimens were sacrificed to the anatomical examination.

According to Dall the color of the living animal is "bluish." The animals preserved in spirits were of a uniform whitish color, so too the rhinophoria and the branchia. Their length was 9–12.0 mm. by a breadth of 5.3–7.0 and a height of 3–4.5 mm.; the breadth of the foremost part of the foot 3.6–5.0 mm. The height of the rhinophoria reached about 2.2 mm., of the branchial leaves 1.0 mm.

The form almost entirely as in the typical form and as in the *L. hystricina*. The head as in the last species; also the openings of the rhinophor-holes, with their (mostly three) larger tubercles, set with equal spaces; the club of the rhinophoria with about twelve to fifteen rather thick leaves. The tubercles of the back as in the *L. hytricina*; the number of larger ones much exceeding that of the smaller, which are scattered between them. The branchial disk as in the *L. hystricina*, also the branchial leaves, whose number did not surpass twelve to twenty. The foot as usual.

The central nervous system (fig. 1) nearly as in the *L. hystricina*. The cerebro-visceral ganglia of roundish or oval form, as also the pedal ones which were not much smaller than the former. The com-

---

[1] In my "Malacolog. Unters." (Semper. Philipp. II, ii, Heft xiv, 1878, p. 606–613; Tab. lxiv, fig. 13, 14–19; Tab. lxv, fig. 1–5, 6–13) I have given some anatomical remarks on the typical *L. bilamellata* and on the Greenlandic variety (*D. liturata*, Beck).

missura pedalia nearly as long as the diameter of the pedal ganglia; the subcerebral lying rather close up to the pedal; the visceral quite free, much thinner. A very short-stalked smaller ganglion (fig. 1c) connected with the under side of the right visceral ganglion, gives off a nerve that swells into a new ganglion, which sends out three nerves (N. genitalis). The olfactory ganglia short-stalked, spindle-shaped. The buccal (fig. 1d) and the gastro-œsophageal ganglia (fig. 1e), nearly as in the *L. hystricina*; the commissure between the first extremely short, the gastro-œsophageal somewhat smaller.

The nervi optici one to one and a-half times as long as the diameter of the cerebral ganglia; the eyes with black pigment, yellowish lens. The otocysts (fig. 1) lying rather backwards, a little smaller than the eyes; the otokonia of the usual form, in number about fifty. The leaves of the rhinophoria without spicula. In the skin were almost no spicula and no larger or calcified ones on the surface of the rigid papillæ of the back, which thus were rather smooth. In the interstitial connective tissue small calcified cells, but no larger spicula.

The mouth-tube as in the *L. hystricina*. The bulbus pharyngeus as in that species, but the sheath of the radula shorter and less prominent, bent upwards, sideways or down and forwards. On the interior part of the nearly colorless labial disk, the usual belt of (about twelve to fifteen) rows of small denticles. The tongue strong, rather long, with curved superior and nearly straight inferior margin. In the mature radula twelve to fourteen or sixteen rows of teeth, further backwards fifteen or sixteen to eighteen rows of developed, and three of partly developed teeth; the total number of rows thus thirty, thirty-one or thirty-five to thirty-seven. The median plates (fig. 14) of nearly the usual form, in the under side rather excavated, with thickened margins. The large lateral plates (fig. 13) formed nearly as in the *L. hystricina*, but larger, reaching a height of 0.12 mm.; the denticulation of the interior margin of the hook stronger, with more (about twenty) denticles and reaching farther out towards the end of the hook. The exterior plates nearly of the same form as in the last species, reaching to the height of about 0.6 mm.

The sucking-crop quite as in the former species.

The salivary glands much smaller than in the former species, reduced to a large, scarcely lobed, whitish mass on each side of the root of the œsophagus.

The œsophagus somewhat spindle-shaped. The stomach included in the liver. The intestine issuing from the liver behind its middle.

The liver of grayish-white color, of the length of about 9.5 mm. by a breadth of 4 and a height of about 3.75 mm.; the hinder end rounded, the fore-end rather truncated, the anterior one-third on the upper and right side flattened by the anterior genital mass.

The heart and the renal syrinx as usual; the median renal chamber continued to the fore-end of the liver. The sanguineous glands connected on the upper side of the central nervous system to a flattened whitish mass.

The glandula hermaphrodisiaca clothing the upper side of the liver, and scarcely distinct from it in color; in its lobules were large oögene cells. The anterior genital mass compressed, plano-convex; 4.0 mm. long, by a height of about 3.3 and a breadth of 1.2 mm. The albuminous gland on the left side of the mass and forwards, yellowish, very finely gyrated on the surface; the mucous gland whitish, pellucid. The spermatoduct as well as the (3.0 mm. long) penis as in the *L. echinata*. The spermatotheca rather small, spherical.

**L. varians, var.**

To this same species belonged certainly five specimens of a *Lamellidoris*, which were taken by Dall in July, 1873, at Unalashka Island (Aleutians), at the depth of sixty fathoms on mud and stones. Nevertheless, the color of these animals in the living state was, according to Dall, "yellowish-white."

The size and the particular measures accorded with those of the more typical individuals, referred to above.

The central nervous system as just mentioned, so even the eyes and the otocysts. The bulbus pharyngeus of the usual form; on the tongue eleven rows of teeth, farther backwards twenty-six developed and four not quite developed rows, the total number thus forty-one. The plates quite as formerly described. The sucking-crop quite as in the typical form, also the salivary glands. The whitish sanguineous gland entirely covering the central nervous system. The penis as usual.

Two specimens of another variety of this form were gotten by Dall, in July, 1873, at Kyska Island, on sandy bottom, and at a depth of nine to fourteen fathoms. In a living state they were, according to Dall, of yellowish color.

The length of the animals preserved in spirits was 8.5 to 9.0 mm., by a breadth of 6.0 mm., and a height of about 3.5 mm. The color was uniformly whitish or yellowish-white. One individual was dissected.

The central nervous system was as above mentioned, and also the eyes (their nervi optici rather long), and the otocysts (the number of the otokonia about one hundred). The bulbus pharyngeus as usual; on the tongue sixteen rows of teeth, farther backwards eighteen rows of developed and four of younger teeth; the total number of rows, thirty-eight. The plates as above; the length of the median plates 0.05 to 0.058 mm.; the height of the anterior large lateral plates about 0.14 mm., of the posterior about 0.17 mm.; the number of denticles on these plates mostly fifteen to twenty. The vesica fellea was at the left side of the pylorus.

3. **L. hystricina**, Bergh.

    *L. hystricina*, Bergh, Mal. Untersuch., l. c., 1878, p. 614, Tab. lxviii, fig. 17-23.

Color cœrulescens.

Dentes laterales margine interno denticulati sed non usque ad apicem.

*Habitat.* Oceanum Pacificum (insula Kyska).

One specimen of this species was found by Dall, at Kyska Island (Aleutians), on rocky bottom, at a depth of ten fathoms, in June, 1873. According to Dall, the color of the living animal is bluish.

The specimen preserved in spirits was 9.5 mm. in length, reached a breadth of 6.0 mm., and a height of the true body (without the papillæ) of 3.5 mm.; the breadth of the foremost part of the foot was 5.3 mm., the height of the rhinophoria was about 2.1 mm., of the branchia about 1.2 mm., of the dorsal papillæ 1.2 mm. The color was uniformly whitish.

The form was oval, the back not very convex. The head rather large, formed like a velum, that is radiately folded, and has its side parts connected with the ends of the anterior margin of the foot; in the middle of the hinder part of the under side of the velum is a transverse slit, in which the small mouth-pore opens. The opening of the rhinophor-holes was nearly round, with the margin rather thin, here were three papillæ of the same kind as on the back; the rhinophoria stout, the club with about twenty leaves. The back covered all over with mostly stout, club-shaped papillæ, apparently set without order, and extending nearly out to the very margin of the mantle, which is thin and has on the upper side smaller, cylindrical or club-shaped papillæ. The papillæ all firmly adherent to the skin, the spicules shining through all over on the back and in the papillæ. The branchial disk rather large, at the margin set with about fourteen papillæ, irregularly alternating in size. The branchia composed of twelve small leaves of the usual kind. The centre of the disk and the anus as usual. The foot somewhat shorter and narrower than the back, broader in front, with the anterior margin rather straight, rounded posteriorly.

The cerebro-visceral ganglia showed the visceral part a little larger than the cerebral, the pedal somewhat smaller than the visceral; the four commissures as usual; the offshoot of the nerva genitalis could not be determined. The buccal ganglia rounded, connected through a short commissure; the gastro-œsophageal having about one-quarter of the size of the latter.

The eyes with very rich black pigment; the nervus opticus not short. The otocysts as large as the eyes, filled with otokonia of the usual kind. In the thin leaves of the rhinophoria no spicula. In the skin of the back and in the dorsal papillæ an enormous amount of irregular or rounded particles, often coalescing together in larger, irregular lumps, which very often were crowded together in irregular heaps; in the papillæ also were long, strong and very much calcified spicula, often of uneven surface, whose points, as usual, often projected on the surface of the papillæ. In the interstitial connective tissue, including the ends of the different ducts of the genital organs (vagina, mucous gland duct), masses of large and long (as much as 0.9 mm.), calcified spicula.

The mouth-tube was about 1.0 mm. long, rather wide, with strong, longitudinal folds. The bulbus pharyngeus of usual, irregular form, the bulbus proper of the length of about 1.75 mm.; the sheath of the radula, nearly as long as the bulbus, curved downwards. The labial disk oval, at the inner margin of darker color, and there showing (fig. 17) a narrow belt of small, yellowish denticles, of a height of 0.007 to 0.015 mm.;[1] this belt seems continued a short space up in the mouth that is otherwise, like the rest of the buccal cavity, clothed with a rather thick, yellowish cuticula. The tongue rather long and narrow, in the groove on its back sixteen rows of teeth, in the sheath eighteen developed and six undeveloped rows, the total number consequently forty. The color of the true lateral teeth yellowish, the others nearly colorless; the height of the outer pseudo-plates about 0.075 mm. The median pseudo-plates elongate, narrow (fig. 21); the true (lateral)

---

[1] In the outer mouth was found a little Caprella, of the length of 3.0 mm.

teeth strong, finely denticulated (with six to eight denticles) on the inner side of the hook, and with a strong, rounded prominence at the base of this (fig. 18a, 19, 20); the external pseudo-plates with the usual curved points (fig. 18b). Irregularities in the form of the last were often observed (fig. 23).[1]

The crop entirely as in the typical species, the largest diameter 1.3 mm.

In the stomach indeterminable animal matter and a little, undeterminable worm, of the length of 2.0 mm.

The hermaphroditic gland as usual; the lobules filled with sperma. The anterior genital mass rather large, measuring in length 4.5 mm., in height 2.5 mm., and in breadth 2.3 mm.; the left side flat or a little excavated, the right rather convex. The mucous gland, as well as the albuminous gland, white and yellowish-white. The spermatoduct not very long, but rather strong, continued in the very strong penis, that (retracted) forms the fore-end of the whole mass. The penis has a length of about 3.5 mm., by a diameter of 1.3 mm.; the inferior end rather constricted; the superior three-quarters of the organ compact, perforated through the axis by the dense coils of the spermatoduct proper; the inferior one third hollow, including the curved and pointed glans.

2. **L. muricata** (Müller). Plate IX, fig. 18; Plate XI, fig. 10-12.

*Doris muricata*, O. F. Müller. Zool. Dan. Fas. III, 1789, p. 7, Tab. LXXXV, f. 2, 3, 4.

*Doris muricata*, Müller. Sars, (forma ?) Lovén, Ind. Moll. Scand. 1846, p 5.

*Doris muricata*, Meyer und Moebius. Fauna der Kieler Bucht, I, 1865, p. 73-75, Taf. Vc, fig. 1-8.

? *Lamellidoris muricata*, Müller. G. O. Sars, Moll. reg. arct. Norv., 1878, p. 307, Tab. XIII, fig. 6.

Color flavidus vel luteo-albus.

Dentes laterales magni hamo denticulato sed non usque ad apicem.

*Hab.* Oc. Atlanticum septentr.

The original specimen on which Müller founded his *Doris muricata* does not exist, and by his incomplete description it is now completely impossible with full certainty to determine what species was meant by his description. In future the species described by Meyer and Moebius

---

[1] From the presence of only one individual, the examination of the radula was extremely difficult and limited, as also that of the genital organs.

and by me ought to be called by that name. To the same is without doubt to be referred the second variety (♂) of the *D. muricata* (Müller, Sars) of Lovén (the first being the *D. Lovéni* of Alder).

Of this form, and under that name, I have had two well conserved specimens for examination, kindly sent me by Mr. Friele, of Bergen, and caught in the neighborhood of that place.

The individuals (preserved in spirits) were of light yellowish color.[1] The length 9–10 mm. by a breadth of 5–6.0 and a height of nearly 5.0 mm.: the breadth of the foot reaching 3.5 mm.; the height of the rhinophoria 1.5, of the branchial leaves 1.0 mm. The form of the animal as usual; the warts of the back not large, mostly truncate, clavate. The openings for the rhinophoria as usual, with two tubercles before them, or one on each side; the club with about fifteen to twenty leaves.[2] The branchial leaves about twelve to fourteen, as far as could be determined;[3] the space inclosed by the gill covered with the usual tubercles; the anus presenting the ordinary features. The head rather large, the side parts adhering to the foot throughout their whole length. The genital groove with three openings; a foremost round, a median spalt-formed, and a posterior large and round.

Both individuals were dissected; the peritoneum was colorless.

In the central nervous system the cerebro-visceral ganglia appeared rather short, reniform; the pedal ones of roundish form, somewhat larger than either of the former; the commissures rather short. The olfactory ganglion short-stalked, nearly spherical, situated rather posteriorly on the upper side of the cerebral ganglia, and nearly as large as the buccal ones. The buccal ganglia of oval outline, connected by a short commissure; the gastro-œsophageal nearly spherical, in size about one-quarter of the former, short-stalked: a secondary ganglion lying above the last on the œsophagus.

The eyes not short-stalked; with rich black pigment and yellow lens. The otocysts a little smaller than the eyes, filled with otokonia of the common kind. In the leaves of the rhinophoria rather few but large spicula of the same kind as in the skin, more or less perpendicular on the free margin; the axes of the club like the stalk still more richly endowed with smaller and larger spicules. Under the glass the

---

[1] According to Lovén the color is yellowish; to Meyer and Moebius white or yellowish-white, the rhinophoria orange-colored.

[2] According to Meyer and Moebius the club of the rhinophoria has but nine or ten leaves.

[3] Meyer and Moebius mention eight leaves as nearly constant.

skin between the warts, as well as the warts themselves, showed the white spicules everywhere shining through; the spicules often projecting from the surface of the warts. The spicules for the greater part very large, long, and reaching a diameter of at least 0.05 mm.; they were strongly calcified, mostly straight or slightly curved, the surface nearly even. In the interstitial tissue were rather many spicules, but (as in the rhinophoria) less calcified than in the skin.

The mouth-tube rather wide. The bulbus pharyngeus of nearly usual form, about 1.6 mm. long; the sheath of the radula, moreover, projecting backwards about 0.4 mm., bent somewhat upwards or downwards; the lip-disk with a rather thick yellowish cuticula; the sucking-crop large, larger than the true bulbus, to which it adheres by a very short petiolus. The tongue with nine rows of teeth, further back twenty to thirty-two developed and three younger rows; the total number of rows, thirty-two to forty-four.[1] The yellow median plates (fig. 10a) about 0.05 mm. long, of the usual form. The large lateral plates yellow, of about 0.12 mm. height; the form as usual; the hook with about fifteen to sixteen fine denticles, and a strong tooth at the inside of the base (fig. 10bb). The external plate colorless, about 0.04 mm. in height, with the usual rudiment of a hook (fig. 10c, 11b).[2]

The salivary glands white, rather thick, making two or three short coils at the sides of the œsophagus. The œsophagus as usual. The intestine emerging from the liver at about the middle of its length; the biliary sac (fig. 18) is at the pyloric part of it, situated deeply, scarcely showing itself on the surface of the liver and opening (fig. 18a) into the stomach close to the pylorus. The liver about 6.5 mm. long by a breadth of 3.0 mm. and a height of 2.0 mm., deeply excavated in the anterior third of its right side, and of light yellow color. The sanguineous gland much flattened, whitish, heart-formed, of about 1.5 mm. largest diameter. The renal chamber rather wide, the tube on its floor strong.

---

[1] Meyer and Moebius (l. c. p. 73) mention twenty-nine rows; Alder and Hancock thirty.

[2] The representations of the external plate by Meyer and Moebius (l. c. fig. 2, 6) are not natural. Alder and Hanc. (l. c., Part VII, p. ii, Pl. 46, supplem. text) mention two external plates in their *D. muricata* (as in their *D. diaphana*); either the *D. muricata* of A. and H. must be another species, or they must have fallen into error from the particular view which is sometimes had in certain positions of the hind ends of the large lateral teeth with the external ones.

The lobes of the hermaphroditic gland without developed sexual elements. The anterior genital mass about 2.5–3.0 mm. in length by a height of 2.0 mm. and a breadth of 1.0–1.5 mm. The ampulla of the hermaphroditic duct of yellowish color, rather thick (—0.75 mm. diameter), making a wide curve. about 2.5 mm. long. The spermatoduct long; its first part thinner, about 9.0 mm. long, then through a stricture of the length of nearly 1 mm., passing into the thicker part, which in its last half increases in thickness, and, all in all, has the length of about 6.0 mm. by a diameter of 0.75 mm.; the last part (fig. 12c) passes into the penis[1], in whose cavity (fig. 12bb) the glans (fig. 12a) projects as a short club, the proper seminal duct passing down to the gland in nearly continual cork-screw windings, and often shining through the walls of the external coat. The spermatotheca whitish, nearly spherical, of about 1.3 mm. diameter, filled with seminal matter and detritus; the spermatocysta elongate, nearly twice as long as the former, yellowish, deeply imbedded in the mucous gland, filled with ripe semen; its duct somewhat longer than the cysta. The vagina short.[2] The mucous gland yellowish and yellow.

The species approaches to the *L. hystricina* and *L. varians* (of the Pacific), but differs entirely in its colors; still the possibility cannot be denied that further investigations may show both the Pacific "species" to be merely varieties of the old *Lamellidoris muricata* of the Atlantic.

## ADALARIA, Bergh.

*Adalaria*, R. Bergh. Malacolog. Unters. (Semper, Philipp. II, ii). Heft XIV, 1878, p. xl.
*Adalaria*, R. Bergh. Gattungen nord. Doriden, l. c. 1879, p. 360.

Forma corporis fere ut in Lamellidoridibus. Nothæum papillulatum vel subgranulosum. Branchia (non retractilis) e foliis vix multis, in formam ferri equini ut plurimum dispositis formata. Caput ut in Lamellidoridibus, latum, semilunare, tentaculis vix ullis vel brevissimis lobiformibus. Aperturæ rhinophoriales integræ, tuberculis anticis 2–3, calvitie postica.

Discus labialis non armatus. Lingua rhachide lamellis depressis instructa; pleuris dente laterali interno hamiformi majore et serie

---

[1] The exserted penis is figured by Meyer and Moebius (l. c. taf. fig. 4) and mentioned as cylindrico-conical.

[2] The upper end of the vagina seemed to present a particular diverticle.

dentium externorum sat applanatorum præditis.   Ingluvies buccalis bulbo pharyngeo petiolo connata.

Penis glande parva inermi.   Vagina brevis.

The genus has been established by the author (1878) to receive the *D. proxima* and its allies. The *Adalariæ* externally approach nearest to the *Lamellidorides*; their branchial leaves are also disposed mostly in horseshoe form, but fewer in number. The head and the tentacles are more as in the *Acanthodorides*. The back is nearly as in the *Lamellidorides*, but the granules are sometimes more pointed. The opening for the rhinophoria as in the *Lamellidorides*, with plain margin; before them two to three tubercles, behind them the glabella. The lip-disk only covered by a strong cuticula. The armature of the tongue approaching to that of the *Acanthodorides*. The rhachis of the tongue carries depressed small yellow plates; at each side of these a large hook-formed yellow plate, and further outwards a series of smaller, nearly colorless plates, of which the inner ones are more compressed, the rest depressed. The sucking-crop as in the *Lamellidorides*, through a petiolus fixed to the bulbus. The salivary glands as in the *Lamellidorides*. The œsophagus wider at its root. The penis unarmed; the vagina short.

The *Adalariæ* are *Lamellidorides* with a tongue resembling that of the *Acanthodorides*; they form a sort of connecting link between these two groups.

Of the typical species, the spawn is known (through Alder and Hancock) and some few notices have been published about their biology (through Meyer and Moebius); Sars mentions[1] the swimming of *Ad. Lov'ni*.

The genus seems to belong to the northern oceans; only five species seem hitherto known.

1. *Ad. proxima* (A. et H.).   Oc. Atlanticus sept.
2. *Ad. pacifica*, Bgh., n. sp.   Oc. Pacif.
3. *Ad. virescens*, Bgh., n. sp.   Oc. Pacif.
4. *Ad. albopapillosa* (Dall).   Oc. Pacif.
5. *Ad. Lovéni* (A. et H.).   Oc. Atlant. sept.

1. **Adalaria proxima** (Alder et Hancock).   Pl. IX. fig. 12-15.

   *Doris proxima*, A. et H.   Monogr. Part VI, 1854.   Fam. 1, Pl. 9, figs. 10-16; Part VII, 1855.   Pl. 46, suppl. f. 8.
   *Doris proxima*, Meyer u. Moebius, Fauna der Kieler Bucht, I, 1865. P. 69-71; taf. V b, fig. 1-3.

[1] Sars, Bidr. til Söedyrenes. Naturhist. 1829, p. 17.

Color flavus vel e rubro flavus.
Dentes laterales (magni) hamo edentulo; externi numero 10.
*Hab.* Oc. Atlant. septentr.

Of this form I have had for examination three specimens of nearly equal size, kindly sent me by Prof. Moebius in Kiel, and caught in the neighborhood of that town.

The individuals were of a uniform whitish color, the liver shining reddish-gray through the foot. Alder and Hancock have already remarked this shining through of the liver. The length was 7.0–8.0 mm., by a breadth of 5.0–5.5, and a height of about 3.5 mm.; the height of the rhinophoria about 1.25, of the branchial leaves 0.75 mm. The form nearly as in the *Ad. pacifica*, also the tubercles (fig. 12) of the back and the surroundings of the rhinophor-holes; the branchial leaves nine to ten in number. The number of branchial leaves according to Alder and Hancock is eleven, according to Meyer and Moebius eight or nine. The rhinophoria with about fifteen to twenty leaves. The lateral parts of the head nearly connate with the foot, and only slight traces of true pointed tentacles. The foot as in the next species.

The three individuals were anatomically examined. The peritoneum colorless.

The central nervous system as in the *Ad. pacifica*, but less depressed. The eyes and otocysts as in that species; the last with about 200 otokonia of very varying diameter, reaching about 0.02 mm. The spicula of the skin as described by English and German authors; a rather large quantity spread in the skin of the head.

The bulbus pharyngeus (with the crop) of the length of about 1.5 mm., by a height of 1.5 and a breadth of 0.8 mm.; the crop making about half of the bulbus; the lip-disk with strong yellowish cuticula; the sheath of the radula a little prominent, bent more or less upwards. The tongue narrow and pointed, with seven to nine rows of teeth, further backwards thirty or thirty-one rows of developed and three of younger teeth; the total number thus amounts to forty or forty-three.[1]

The teeth as in the *Ad. pacifica*. The large lateral yellowish, the rest nearly colorless. The length of the median teeth about 0.025 to 0.03 mm. The large lateral (fig. 13bb, 14) showed the prominence

---

[1] Alder and Hancock notice forty-one, Meyer and Moebius thirty-nine rows of plates.

at the inside of the root of the hook quite as in the *Ad. pacifica*. The external teeth (fig. 15) only nine or ten in number,[1] fewer than in that species, always absent on more than half the tongue.

The salivary glands as in the next species, also the œsophagus, the stomach and the intestine. The liver also of nearly the same form, the inferior part of the posterior end continued as a little cone; the surface (especially of the back part) yellowish-white; the substance yellow. The vesica fellea in its usual place, small. The heart as usual, also the sanguineous gland. The renal syrinx and the urinary chamber as usual.

The anterior genital mass rather compressed, of angular-roundish outline, of about 1.75 mm. largest diameter. The spermatoduct seemed shorter than in the next species, especially the second part; the penis short. The spermatotheca pyriform; the spermatocysta of more oval form, having only about one-quarter of the size of the former, and filled with sperma. The mucous gland whitish and yellowish.

2. **Adalaria pacifica**, Bergh, n. sp., Pl. IX, fig. 17; Pl. X, fig. 1-3; Pl. XI, fig. 15.

Color lutescens.

Dentes laterales (magni) hamo edentulo; externi numero 15.

*Habitat.* Oceanum Pacificum (Unalashka).

Of this species Dall caught three specimens, in September, 1874, at Unalashka, on a bottom of mud and shells.

According to Dall, the color of the living animal is "yellowish;" the specimens preserved in spirits were of a uniform yellowish color. The length of the two larger specimens about 12.0 to 14.0 mm., by a breadth of 8.0 to 9.0 mm., and a height reaching 4.5 to 5.0 mm.; the breadth of the foot 6.0 mm., the height of the rhinophoria about 1.5 mm., of the branchial leaves 1.2 mm.

The form as in the *Ad. proxima*, a little broader anteriorly. The back covered all over with a mass of rather stout, subglobose and subpetiolate tubercles quite as in the typical species, mixed with much fewer smaller ones. The larger ones, under magnification, showing the perpendicular spicula shining through, while other spicula were detected irregularly scattered in the intervals between the tubercles. The rhinophor-holes nearly without projecting margin; the adjoining part of the back, behind, smooth; immediately before the holes, on

---

[1] The number of external plates is, according to Alder and Hancock, ten, to Meyer and Moebius, eight or nine.

the contrary, two or three larger tubercles; the club of the rhinophoria with about thirty leaves. The branchial area surrounded by larger tubercles. The branchial leaves in number, eleven or twelve; immediately before the two hindermost was the slightly prominent anus, and at its right side the renal pore; in the space between the anus and the branchial leaves, three or four larger and two or three smaller tubercles. The head large; the tentacles short, pointed. The foot broad, rounded behind, a little broader in front; the furrow on the anterior margin very indistinct. The three individuals were all dissected. The peritoneum was colorless.

The central nervous system rather flattened; the cerebral ganglia larger than the visceral, which were lying at their outer margin and were a little larger than the pedal ones; the proximal olfactory ganglia bulbiform, less large than the buccal ones, which were of short, oval form, connected through a very short commissure; the gastro-œsophageal ganglia short-stalked, rounded, nearly half as large as the former, with a very large cell. The subcerebral and the pedal commissures connected, the visceral free.

The eyes with coal-black pigment, yellow lens; the nervus opticus in one individual with black pigment. The otocysts, under a magnifier, very distinct as chalk-white points at the hinder margin of the cerebral ganglia, nearly as large as the eyes, filled with ordinary otokonia. In the leaves of the rhinophoria scanty, scattered spicules, perpendicular on the free margin, not much more calcified than in the skin; in the stalk of the organ the spicules larger and less scanty. The skin, especially its tubercles, with many long spicules and calcified cells and groups of such cells; the form of the spicules different from that of the *Doris proxima*, as figured by Alder and Hancock (Monogr., Part vi, fam. 1, Pl. 9, fig. 15), and by Meyer and Moebius (l. c., figs. 8, 9), much less calcified, more straight and of more uniform shape. In the interstitial connective tissue of the chief ducts of the anterior genital mass were scattered large spicules.

The mouth-tube wide, about 1.3 mm. long. The bulbus pharyngeus of rather compressed form, about 2.0 mm. long; the sheath of the radula strongly projecting from the hinder end, nearly as long as the bulbus, more or less curved upwards; the lip-disk oval, with a very strong yellowish cuticula. The tongue with ten or eleven rows of plates, further back twenty-nine to thirty-four rows of developed and three of younger plates; the total number thus forty-two, forty-three, forty-seven. The median plate (Pl. IX, fig. 17*a*; Pl. X, fig. 1) yellowish, of a length of about 0.045 mm., with a median furrow along the upper side and with thickened margins. The large laterals horn-yellow in color, reaching the height of about 0.1 mm. (Pl. IX, fig. 17$b$; Pl. X, fig. 2$aa'$), hook-shaped, with a strong, rounded prominence at the inside of the root of the hook (fig. 17). On each side (Pl. X, fig. 2$b$, $c$) of the two large plates (in two individuals) constantly fifteen smaller, nearly colorless plates of a length of about 0.06 mm. These plates were all somewhat depressed; the five inner ones smaller, somewhat compressed (fig. 2, 3$a$, 15); the others (fig. 2, 3) broader, with the upper edge broad and irregularly toothed; the outermost (fig. 2$c$) a little smaller than the adjoining plates. The bases in each of these (fifteen) plates large, forming nearly half, or at least making more than a third of the size of the whole plate.[1] The crop of the bulbus of the usual form, as large or a little larger than the bulbus itself; with a very short stalk with strong longitudinal musculature, its aperture opening immediately behind the lip-disk.

The salivary glands large, white, very elongate, in their foremost part broader, and with several coils filling the space left between the crop, the bulbus and the œsophagus.

The œsophagus long. The stomach small, enclosed in the liver; the intestine rather short, forming its knee behind the fore-end of the liver. The large posterior visceral mass about 9.0 mm. long by a breadth of 4.3 and a height of 3.5 mm.; the posterior end somewhat pointed, though rounded; the fore-end broader, perpendicular, somewhat flattened on the right side; the color of the surface (hermaphroditic gland) ash-gray, the interior (the liver) brown or black brown, or quite yellow.

The heart as usual. The sanguineous gland irregularly reniform, situated somewhat more towards the left side, rather thick, whitish, covering the central nervous system and a large part of the bulbus pharyngeus from above. The renal syrinx as usual.

The hermaphroditic gland without developed sexual elements. The anterior genital mass proper rather small, compressed, of about 2.5 mm. largest diameter, but the loop of the spermatoduct (and the penis) nearly as large as the rest of the mass. The spermatoduct long, in its first part white, rather strong; nearly as long as the second in which it passes through a stricture; this last part is thicker, cylindrical, elongated, about 5.0 mm. long, passing without exact limits into the

---

[1] In both individuals the three to five foremost rows were without the smaller plates, and the following two or three very incomplete in this respect.

short penis. The spermatotheca pyriform, about 1.3 mm. long; the spermatocysta not having one-fourth of the size of the last; both empty. The mucous gland whitish and yellow-whitish.

This seems even externally to differ somewhat from the typical form, of which it nevertheless may prove to be but a variety. Neither Alder and Hancock, nor Meyer and Moebius saw more than eight to (nine) ten external plates on the tongue of *Ad. proxima*, while this Pacific form always presented fifteen.

### 3. Adalaria virescens, Bgh., n. sp. Plate X, fig. 4, 5.

Color virescens.
Dentes laterales (magni) hamo edentulo; externi numero 15.
*Hab.* Oc. Pacific. septentr. Unalashka.

Of this species Dall found four specimens at Unalashka, on gravel, in a depth of nine to fifteen fathoms, in September, 1874.

According to Dall the color of the living animal was "greenish," and the animals preserved in spirits showed remains of the same color as a uniform grayish green. The length of these was 11.5–12.0 mm., by a breadth of 8.0 mm. and a height of 5.0 mm,; the height of the rhinophoria about 2.0, of the branchial leaves about 1.0 mm.

The form, as well as the rhinophor-openings, were quite as usual; the club of the rhinophoria with about thirty-five leaves. The gill not large, with nine to twelve leaves; the space within the gill as usual, also the arms and the renal pore. The back covered with granulations and short clubs. The head, with the tentacula and the genital opening as usual.

Three individuals were dissected; the peritoneum was colorless.

The central nervous system showed the cerebral ganglia larger than the visceral, which were lying on the outside of and behind the former, very distinct from them; the pedal ones being intermediate in size between the cerebral and the visceral ganglia. On the exterior part of each cerebral ganglion a little short-stalked ganglion (gang. opticum?) was easily visible under a hand magnifier. The (proximal) olfactory ganglia bulbiform, short-stalked, a little larger than the buccal ganglia, which were short-oval, connected through a very short commissure; the gastro-œsophageal being about one-fourth to one-fifth of the size of the former. In the neighborhood of the penis a little oval ganglion (g. penis) having a largest diameter of about 0.25 mm. (fig. 5), containing only rather small cells.

The eyes with black pigment; the otocysts with not very many and not much calcified otokonia. No distal olfactory ganglion, as far as could be seen; no spicula in the leaves of the rhinophoria. The skin as in other species; the spicula projecting on the surface of the granulations of the back.

The bulbus pharyngeus about 1–1.5 mm. in length; the sheath of the radula projecting 0.75–1.0 mm., bent upwards; the sucking-crop a little larger than the bulbus itself, short-stalked; the lip-disk as usual. The tongue compressed, rather prominent, with six, eight, and nine rows of teeth, also further back twenty-four, thirty-two and thirty-three developed and three younger rows; the total number of rows thus being thirty-five, forty-one, forty-five. The median plates, the large lateral and the (fifteen) external ones scarcely different from those of the last species.

The salivary glands rather strong, with two or three short coils filling the space at the sides of the œsophagus (fig. 4), white. The œsophagus (fig. 4a) wide in its upper part, the rest narrow. The anteriorly proceeding part of the intestine 2.0 mm. long, the other retroceding part 8.0 mm. long; no biliary sac could be found either at the pylorus or higher up. The liver about 9.0 mm. long by a breadth of 4.2 and a height of 4.0 mm.; of brownish-gray color; the anterior end truncate, inclined downwards and backwards; the anterior one-third of the right side flattened for the anterior genital mass; the posterior end somewhat pointed, rounded at the point.

The sanguineous gland whitish, covering the anterior end of the bulbus pharyngeus and the foremost part of the central nervous system or this last and the hinder part of the bulbus.

The anterior genital mass about 3.5 mm. long by a breadth of 0.75 and a height of 1.5 mm., a very large part of it formed by the thick part of the spermatoduct. The ampulla of the hermaphroditic duct about 2.0 mm. long, rather thin, whitish. The spermatoduct long; the first part thinner, about 8.0 mm. long; the rest making a large curve, about 5.5 mm. long, about three times as thick as the first, with a diameter of 0.6 mm.; the spermatoduct proper making many coils in its interior course downwards to the penis, which shows a little unarmed glans in the bottom of its orifice; in one individual the penis was exserted as a conical prominence of the height of 1.0 mm. The spermatotheca pyriform, about 1.0 mm. long, of grayish color; the spermatocysta a little less large, spherical; the vagina rather short. The mucous gland rather small.

Even this species might perhaps be merely a variety of the former; still it is of a quite different color and the back much more coarsely granulated.

4 **Adalaria albopapillosa** (Dall). Pl. IX, fig. 16; Pl. X, fig. 9-11.

*Alderia* (? ?) *albopapillosa*, Dall, Amer. Journ. of Conch., vii, 2, 1872, p. 137.

Color pallide flavescens, papillis dorsalibus niveis.
Dentes laterales (magni) hamo basi denticulato.
*Habitat.* Oceanum Pacificum septentrion. (Sitka).

Of this curious animal Dall caught three specimens [in company with the *Doris* (*Archidoris*) *Montereyensis* and the *Æolidia* (*Hermissenda*) *opalescens*], in July, 1865, on algæ, at the depth of six fathoms, at Sitka (Alaska).

According to the drawings of Dall, the color of the living animal is very pale yellow,[1] the back all over covered with chalk-white papillæ; the length was 3, the breadth 2 lines. The three original specimens preserved in spirits were of a length of 5.5 to 7.0 mm., of a greatest breadth of 4.0 to 4.5 mm., and a height of 2.75 mm. The color was uniformly translucent grayish and yellowish-whitish. The form of the animal was oval, the mantle a little larger than and hiding the rest of the body. The back convex, covered all over with a multitude of cylindrical or fusiform, relatively rather large papillæ, reaching to the height of a full millimetre, and with some few small ones spread between them. The rhinophor-openings at their usual place, having, as usual (with retracted organs), thin margins; before them always two larger papillæ, behind them a little naked space.[2] The club of the (yellowish) rhinophoria with about twenty-five leaves. The gill rather small; the branchial leaves (yellowish), as usual, set in horseshoe form, lower or at least not higher than the dorsal papillæ, in number, ten to twelve; the anal papilla rather low, with one of the ordinary papillæ before and one behind it; the space between the

---

[1] "Of an opaque white, the remainder of the animal, except the eyes, being translucent yellowish."—DALL.

[2] Dall did not detect the retracted rhinophoria ("tentacles none"); the "black eyes sessile on the anterior surface of the body, near the mantle margin," did not exist in the figure, but in one individual two black sand-particles were lying there. The true eyes of the animal could not be detected through the skin, and were lying more backwards.

branchial leaves and the anus otherwise naked.[1] The genital opening as usual. The foot rather large, with a very fine furrow in the anterior margin. The head as usual; the tentacles relatively rather large.

The three individuals were dissected. The peritoneum was colorless.

The central nervous system quite as in the former species, the visceral ganglions lying outside of the cerebral; no distal olfactory ganglion could be detected; the buccal ganglia connected through a commissure at least as long as the diameter of the ganglion; the gastro-œsophageal ganglia and the eyes as in the former species. The otocysts could not be detected. In the leaves of the rhinophoria the spicula much more scanty. In the skin the same kind of not much calcified spicula as in the former species; the papillæ of the back very richly endowed with such, and commonly with a mass of them projecting with their points (Pl. IX, fig. 16) on the surface of the papillæ.

The bulbus pharyngeus as in the former species; the length about 1.5 mm., two-fifths of which is the straight, backwards projecting sheath of the radula; the cuticula of the lip-disk as usual; the buccal crop somewhat compressed, with rather long pedicel. The tongue with nine or ten rows of plates, farther backwards sixteen or seventeen developed and three younger rows; the total number of them, twenty-nine or thirty. The median plates (fig. 9a, 10a) nearly as in the former species, or a little shorter. The large lateral plates (fig. 9b, 10b) rising to the height of 0.12 mm., yellow; their form as in the former species, but at the inside of the hook at its root were three to six or seven to eight small denticles. The external lateral plates (fig. 10cd, 11) farther backwards, in number constantly eight; the outermost (fig. 11a) very small, the others as in the former species.

The salivary glands, as far as could be determined, were as in the last species; so also the œsophagus and crop; also the stomach and the intestine, which seemed to have the usual bag (pancreas, biliary sac) at the pyloric part. The sanguineous gland flattened, grayish, cordate. The liver of brownish-gray color.

In the hermaphroditic gland no ripe elements were found, and the anterior genital mass was very small.

---

[1] According to Dall, the "anus is terminal under the edge of the mantle." This was erroneous. He did not see the gill, but regarded the dorsal papillæ as "branchial appendages."

The species is easy to distinguish from the former, by its color and especially by the denticulated hook of the large lateral plates.

5. **Adalaria Lovéni** (Alder et Hancock). Pl. X, fig. 6-8.

*Doris muricata?* O. F. Müller, Sars, Bidr. til Söedyrenes Naturh., 1829, p. 15. Tab. II, fig. 7, 8.
*Doris Lovéni*, Alder et Hanc. Ann. Mag. Nat. Hist., 3 Ser., X, 1862, p. 262.
*Lamellidoris Lovéni*, Friele et Arm. Hansen, l. c. p. 3.
*Lamellidoris Lovéni*, G. O. Sars. Moll. reg. arct. Norv., 1878, p. 364. Tab. XIV, fig. 1.
? *Lamellidoris muricata* (Müll.) Abildgaard. Mörch, Faunula Moll. Ins. Färöens. Naturh. Foren. Vidsk. Meddel., 1867, p. 75.[1]
*Doris muricata*, Müller, Sars (σ), Lovén, Ind. Moll., 1846, p. 5.
*Doris muricata*, M. Sars. Reise i Lofoten og Finmarken, 1851, p. 75.

Color dorsi et rhinophoriarum e brunneo lutescens, paginæ inferioris et branchiæ lutescens.

Dentes laterales (magni) hamo edentulo; externi (linguæ) numero 12.

*Hab.* Oc. Atlant. septentr.

This species was first noticed by Sars, who hesitatingly regarded it as perhaps the *Doris muricata* of Mueller. It is, moreover, the principal form of the *Doris muricata* ("Mueller, Sars") of Lovén (his second variety being the true *L. muricata*); has been established (1862) as a species by Alder and Hancock, and has as such been adopted by Friele and Hansen, as well as by G. O. Sars, who lately gave figures of the teeth on the tongue. The species has been much confounded with the "*D. muricata*," which is a *Lamellidoris*; it is certainly distinct from the *Ad. proxima*, and seems also to differ from the other described species.

Of this form I have had fifteen individuals for examination, kindly sent me by Mr. Friele, of Bergen, and dredged in the neighborhood of that place.

[1] According to Mörch (Rink. Grönland, I, 1857. Tillæg 4, p. 78), the *D. muricata*, Sars, should be the *D. liturata*, Beck; this last is a mere variety of the *Lamellidoris bilamellata*, and with this should, on the other hand, according to Mörch (Faunula Mollusc. Isl. Naturh. Foren. Vidensk. Meddel., 1868, p. 203), the *D. proxima* of Meyer and Moebius be synonymous, which belongs to the quite different genus, *Adalaria*. An example more— if such were needed—of the way in which the Nudibranchiata have been synonymized and systematized.

The color of the animals preserved in spirits was uniformly yellowish. The length was 13-15.0 mm., by a breadth of 8.5-9.5 and a height of 4-5.0 mm.; the breadth of the foot 6 mm.; the height of the rhinophoria about 2.5 mm., of the branchial leaves 1.0-1.3 mm.; according to M. Sars the height of the rhinophoria is four to five times that of the tubercles of the back, (l. c. p. 16, also in one of his figures fig. 7). The form as usual; the back covered all over with large rounded tubercles, which rose to the height of 1.5 mm., and were of a similar breadth; they were sessile or more or less subpedunculate, sometimes set in indistinct longitudinal rows; between the larger tubercles everywhere were smaller ones of different sizes; on the margin of the back were tubercles of middle size or smaller; the spicula rather indistinct between and in the tubercles. The rhinophor-openings as usual, two large tubercles before them; the club of the organs with about twenty-five leaves. The gill with eight to twelve leaves; according to M. Sars, the number of branchial leaves is ten— to Lovén, eight to ten. A large (high) tubercle between the hindermost leaves, before it the low anal papilla, and to the right side the renal pore; some few smaller papillæ were spread over the space between the anus and the branchial leaves. The head large, broad; the short tentaculæ pointed. The genital opening as usual.

Six individuals were dissected. The peritoneum was colorless.

The central nervous system rather flattened, especially the visceral ganglia, which lay on the outer side of and behind the cerebral ones, which were a little larger; the pedal ones larger than either of the other ganglia, situated perpendicularly upon the former. The proximal olfactory ganglia bulbiform, a little smaller than the buccal ones; no distal could be found. The length of the commissures equal to the largest diameter of the pedal ganglia; the subcerebro-pedal about three times as thick as the visceral. The buccal ganglia of oval form, connected through a short commissure; the gastro-œsophageal about one-sixth of the former in size, with one very large cell.

The eyes with black pigment, yellow lens; the nervus opticus about as long as the largest diameter of the cerebral ganglion. The otocysts of the same size as the eyes, situated externally at the junction of the cerebral and the visceral ganglia; with about fifty ordinary otokonia, but among them four to six larger ones, of a diameter of about 0.025 mm. The leaves of the rhinophoria nearly without spicula; in the axes, and especially in the stalks, on the contrary, an enormous quantity of large spicula, in great part transversely situated. In the skin a rather large quantity of spicula. The broad centres of the warts of the back chalk-white in transverse section, on account of the mass of strong spicula which ascend in bundles through the axes of the warts, their peripheral parts being free from spicula. The spicula, for the most part, staff-shaped or cruciate, reaching a diameter of about 0.08 mm.; small and large rounded ones were also very common; the spicula mostly very strongly calcified. In the interstitial tissue calcified cells were seen scantily.

The mouth-tube was 1.5 mm. long; the bulbus pharyngeus about 1.5 mm. long, the sheath of the radula projecting about 0.75 mm., bent upwards; the sucking-crop nearly as large as the proper bulbus, short-stalked. The lip-disk with the cuticula rather thick, especially at the inferior median line, here sometimes prominent and reminding one of the two blades in the *Acanthodorides*. The tongue (in the six individuals examined) with seven to nine rows of teeth; further backwards twenty-nine, thirty-one, or thirty-four (in three individuals) developed, and three younger rows; the total number of rows was thus forty-two to forty-six. The median plates (fig. 8*a*) and the large lateral (fig. 6*aa*, 7, 8*b*) ones quite as in the *Ad. Pacifica*, also the external ones (fig. 6*b*, 8*c*), but the number of those never surpassed ten or twelve;[1] frequently all gone from the tongue, and only existing in the two to four posterior rows; the height of the large lateral plates rising to about 0.09 mm.

The salivary glands, as usual, white. The œsophagus somewhat wider in its first part; the stomach as usual; the liver of usual form, its substance of yellow color; on the first quarter of the right side an impression for the anterior genital mass. The vesica fellea rather smaller, on the right side of and a little behind the pyloric part of the intestine, with its upper end appearing on the surface of the liver; the duct nearly as long as the bag, opening in the stomach.

The sanguineous gland of subquadratic form, the largest diameter about 2.3 mm., very much flattened, whitish. The tube on the floor of the renal chamber rather strong.

The hermaphroditic gland clothing the liver with a thin, whitish-gray layer. The anterior genital mass small, nearly undeveloped, much compressed, of about 1.75 mm. in length, the height a little less. The ampulla of the hermaphroditic gland thin, otherwise as usual,

---

[1] According to Friele and Hansen (l. c. p. 3) the number of external plates is twelve; the figure of these authors (Tab. II, fig. 1) is rather bad. G. O. Sars has eleven to twelve external plates in his figure.

The spermatoduct as usual, also the penis.[1] The spermatotheca and the spermatocysta as usual. The mucous gland very small, whitish and yellow.

## ACANTHODORIS, Gray.

*Acanthodoris*, Gray, Figs. of Moll. Animals, iv, 1850, p. 103, Guide Moll. Brit. Mus. 1857, p. 207.

*Acanthodoris*, Alder and Hancock, Mon. Brit. Nud. Moll., vii, 1855, p. 43, app. p. xvii. G. O. Sars, Moll. reg. arct. Norvegiæ, 1878, p. 398, Tab. xiv, fig. 4.

*Acanthodoris*, R. Bergh, Gattung. Nord. Doriden, l. c., 1879, p. 356-360.

Forma corporis subdepressa. Nothæum supra sat grosse villosum. Branchia (non retractilis) e foliis tripinnatis non multis et in orbem positis formata.

Caput latum, veliforme; tentaculis brevibus, lobiformibus. Margo aperturæum rhinophorialium lobatus.

Discus labialis armatura e hamulis minutis formata et infra cuticula incrassata prominenti instructus. Lingua rhachide nuda; pleuris angustis dente laterali, hamiformi permagno et dentibus externis minutis (4–8).

Ingluvies buccalis bulbo pharyngeo connata.

Penis armatura e hamulis minutis formata instructus. Vagina longissima.

The genus *Acanthodoris* was established by Gray, to receive the *Doris pilosa* with its non-retractile gill. Alder and Hancock adopted the genus, made an anatomical examination of the typical form and gave it natural characters, which were then adopted by Gray. In several new malacological publications of a systematic nature the genus has been omitted, and in the last twenty years no new information has been published, until G. O. Sars lately gave some notes on the bulbus pharyngeus.

The *Acanthodorides* approach the *Lamellidorides*, yet differ externally in the scattered soft villosities of the back and in the smaller number of the leaves of the gill, which are arranged in a circle.

Internally they differ still more, in the presence of a strong, oral armature, in a different dentition (4+8+1+0+1+8+4), by a pecu-

---

[1] Sars (l. c. p. 16) mentions and figures (fig. 8) the penis as "a large, white, conical" organ.

liarly armed penis and by the imbedding in the pharyngeal bulbus of the buccal crop.[1]

The *Acanthodorides* are not much depressed. The back is covered with soft villi or papillæ; the openings for the rhinophoria have lobed margins. The gill is not retractile, and consists of several (generally seven to nine) tripinnate leaves, quite distinct from one another.[2]

The labial disk is provided with a densely set armature of small hooks, passing backward on the cuticula of the mouth. This last also, in the lowest part of the mouth, at each side of the median line is thickened and projects like two thin, lancet-shaped blades over the bare space left between the lower parts of the prehensile collar.[3] The form of the bulbus pharyngeus is as in the *Lamellidorides*, but the buccal crop is imbedded in the upper wall of the bulbus, opening into it through a slit, and is not conected with it by a short stalk.

The tongue is not broad, but nearly fills the buccal cavity, with a flat furrow for the radula. This last has a naked rhachis, with a low and narrow, longitudinal fold. The pleuræ contain a very large, compressed, upright, lateral plate, with a large body and a rather short, strong hook, denticulated or plain along the inner margin; at the outer side of the large plate are several (four to eight) small, external plates (increasing in number backwards). The salivary glands long, thicker in their foremost part. The œsophagus with a little, crop-like diverticle at its root. Above the pyloric part of the intestine opens a

---

[1] The genus *Calycidoris*, of Abraham (Notes on some new genera of Nudibranchiate Moll., Ann. Mag. Nat. Hist., 4th ser., xviii, 1876, p. 132; and Revision of the Anthobranchiate Nudibr. Moll., P. Z. S., 1877, p. 224), which is said to be allied to the *Acanthodorides* and *Lamellidorides*, still differs by its "subretractile" gill, with simple pinnate leaves, and does not possess external plates on the radula. The genus is very probably apocryphal; in the phanerobranchiate *Dorididæ* it often happens that the gill appears as if more or less retracted in a cavity. A single new species is mentioned, of unknown habitat, the *C. Guntheri*, Abr., l. c., p. 133, Pl. vi, fig. 1.

[2] Alder and Hancock mention and figure (l. c., Pl. 15, fig. 2, 3) the branchial leaves as "united at the base;" so do Meyer and Moebius (l. c., p. 65); this is not the case. The leaves are quite isolated, but there are usually one or two foliola standing between them, which might simulate a coherence of the leaves (cf. also Pl. xv, fig. 6, A. and B.).

[3] These thickenings of the cuticle have been regarded, both by Alder and Hancock, and more lately by Meyer and Moebius (l. c., p. 64, taf. v A, fig. 8, K 9), as "jaws," but have hardly anything in common with those organs properly so called.

little sac, which seems to be homologous with the biliary sac (pancreas, autt.) of other *Doridadæ*. Alder and Hancock, therefore, have denominated that part of the digestive tract as "stomach," although it in no essential respect differs from the rest of the intestine, and is just like that part in the *Chromodorides*, and should be undoubtedly regarded as the pyloric part of the intestine, when that sac opened lower down, as in the *Chromodorides*,[1] in the cavity, which is included in the liver, and seems to be the true stomach. The spermatoduct and the chief duct of the spermatotheca (vagina) are of very considerable length; the former consisting of two different parts, a superior softer, and an inferior very muscular part, internally clothed with an armature, which is continuous through the penis. This last is rather short, the superior part solid and projecting as an armed glans into the inferior, hollow part (præputium). The armature consists of rows of hooks continued in the interior of the organ, and, as mentioned above, farther upwards; quite like that of the *Polyceridæ*,[2] *Phyllidiida*[3] and *Doriopsidæ*.[4]

About the biological relations of these forms very little is yet known and that only with reference to the typical species, through Alder and Hancock, as well as Meyer and Moebius. The spawn is figured by Alder and Hancock l. c., Pl. 15, fig. 9, and by Meyer and Moebius (l. c., fig. 13, 14); about the development nothing is yet known.

The few known species of this genus seem limited to the northern parts of the Atlantic and of the Pacific.

1. *Acanthodoris pilosa* (O. F. Müller).   Oceanum Atlanticum et Pacificum.
    *Doris pilosa*, Cuv.
    *Doris stellata* (Gm.), Cuv.[5]

[1] Cf. my Malacolog. Unters. (Semper, Philipp., II, ii), Heft xi, 1877, p. 464-494; Neue Nacktschnecken der Südsee, ii, Journ. der Mus. Godeffroy, Heft viii, 1875, p. 72-82; idem, iv, l. c., Heft xiv, 1879, p. 1-21.

[2] Cf. my Malacolog. Unters. (Semper, Philipp., II, ii), Heft xi, 1877 (Trevelyana, Nembrotha).

[3] Cf. my Bidr. til en Monogr. af Phyllidierne. Naturh., Tidskr. 3, R. V., 1869; Malacolog. Unters. (Semper, Philipp., II, ii), Heft x, 1876, p. 377-387.

[4] Cf., l. c., Heft x, 1876, p. 384-387; Journ. der Mus. Godeffroy, Heft viii, 1875, p. 82-94.

[5] According to Fischer (Note sur quelques espèces du G. *Doris*, décrites par Cuvier, Journ de Conchyl. 3 sér. x, 1870, p. 290, the *Doris stellata*, Cuv., and the *D. lævis*, Cuv., are identical with his *D. pilosa*, and this with the typical form of Müller.

The *D. stellata* of Philippi seems a quite different form, a *Platydoris*

*Doris lævis*, Cuv.
? *Doris fusca*, O. F. Müll., Zool. Dan. (descr.).[1]
? *Doris tomentosa*, Lovén, Index Moll. 1846, p. 4.

2. *A. subquadrata* (Ald. et Hanc.). Oceanum Atlanticum.
*Doris subquadrata*, A. et H. Monogr., Part. V, 1851, fam. 1, Plate 16, f. 1-3; Part VII, 1855, p. 43, and III, Pl. 46, Suppl. f. 14.
? (*D. stellata*, Cuv. ?). Lebert, Ecob. über die Mundung einiger Gasteropoden. J. Müller, Arch., 1846, p. 441-446, Taf. XII, fig. 10-13.[2]

3. *A. cærulescens*, Bgh., n. sp. Oceanum Pacificum.

4. *A. ornata*, Verrill. Notice of recent additions to the mar. fauna of the eastern coast of North Amer. XXXVIII; Amer. Journ. of Sc. and Arts, XVI, 1878, p. 313. Oc. Atlant.

5. *A. stellata* (Gm.), Verr, l. c., p. 313, *D. bifida*, Verr. Oc. Atlant.

6. *A. citrina*, Verr., l. c., p. 313. Oc. Atlant.

7. *A. ? mollicella*, Abraham, l. c., 1877, p. 228, Pl. XXX, fig. 1-4. Oc. Pacificum.

8. *A. ? globosa*, Abr., l. c., 1877, p. 228, Pl. XXX, fig. 5-9. Oc. Pacif.

1. **Acanthodoris pilosa** (O. F. Müller). Plate X, fig. 12-15; Plate XI, fig. 1-2; Plate XII; Plate XIII, fig. 2-5.

*Acanthodoris pilosa* (O. F. Müller), Alder and Hancock. Monogr. Br. Nudibr. Moll., Part V, 1851, fam. 1, Plate I, f. 1, 3-5, 12; Plate 2, f. 2-6; Plate 15; Part VII, 1855, Plate 46; Suppl. Plate 48, f. 1.
*Doris pilosa* (O. F. Müller), Meyer und Moebius, Fauna der Kieler Bucht, I, 1865, p. 63-67 c. tab.; taf. V, A.

Color paginæ superioris corporis albus vel luteus vel fuscus vel griseus vel rubro-brunneus vel niger.

Dentes radulæ hamo pro parte denticulato.

*Hab.* Oceanum Atlanticum septentr., Pacific. septentr.

(*Platyd. Philippii*, Bgh.). Cf. my Malacolog. Untersuch. (Semper, Philipp. II, ii.). Heft. xii, 1877, p. 507.

[1] It is in most cases a quite useless task to try to elucidate the species of Dorides of the elder authors; their examinations were all too superficial and their descriptions don't contain the data necessary for their verification. The best way would be to wholly cancel these names (*D. fusca*, M.; *D. lævis*, L., etc.) which have given later authors so much trouble. On the *Doris fusca* of O. Fabricius, Mörch has even formed a genus *Proctaporia* (Rink. Grönland. I, 1857. Tillag. 4, p. 78), that must be cancelled, too.

[2] The short statements of Lebert about form and color of the animal examined by him can scarcely entirely prohibit the identification of it with the species described by Alder and Hancock. The figures of the (tongue) teeth given by Lebert, rough as they are, suffice, on the other hand, to secure the identification with the *D. subquadrata*, or at least with a nearly related species.

Of this species I have had a lot of specimens for examination, all preserved in spirits; partly (two) from the neighborhood of Bergen (Norway), kindly sent by Mr. Friele, partly (one) from the Frith of Kiel, sent by Prof. Mo-bius; but particularly (seventeen) from the coast of Denmark (Strib, lille Balt.)

The individuals varied much in color. The variability of the color is noted by Alder and Hancock. They were whitish, or whitish sprinkled with brownish, or dark (bluish) gray, or yellowish, or brownish, or reddish-brown on the back, with whitish or yellowish sides and foot. The length reaching 12.0 mm., by a breadth of 8.0 and a height of 5.0 mm.; the foot then about 4.0 mm. broad, the branchial leaves reaching to the height of about 1.0 mm.

The back covered all over with the soft, slender, conical and pointed, erect (or curved) papillæ of very different sizes, most of them small; between these are larger ones;[1] some of the largest divided into two or three points, and some of them connate and forming small crests, divided above into two or three points. The margins of the sheaths of the rhinophoria rather prominent, divided into several (six to eight) smaller and larger pointed lobes; the club of the rhinophoria with about twelve to twenty leaves.[2] The branchia, in both Norwegian specimens, with eight tripinnate leaves, otherwise with seven to nine (as mentioned by Meyer and Moebius). The anal papilla low, with several papillulæ and a star-shaped aperture; on a low crest, issuing from its posterior, is a strong papilla. The head and the tentacles (Plate X, fig. 14b) as figured by Alder and Hancock (l. c., Plate 15, fig. 1). The anterior margin of the foot with a fine transverse furrow (Plate X, fig. 14a). The genital opening is a longitudinal slit (Plate XI, fig. 2).

The peritoneum was mostly of reddish-brown color.

The central nervous system showed[3] the cerebral ganglia rounded-triangular, not much flattened, a little larger than the more rounded visceral, which lie behind and on the outside of them and show a slight notch in the outside; on the inferior side of the visceral ganglia the pedal ones are set nearly perpendicular on the latter, connected by the

[1] Alder and Hancock, also Meyer and Moebius give eighteen to twenty leaves. Cf. the figures 7-8 of Meyer and Moebius.

[2] Collingwood (Ann. Mag. N. H., 3 ser. vi, 1859, p. 463) remarks that it "when not in motion bears a great resemblance to a miniature hedgehog."

[3] The representation of the system given by Hancock and Embleton On the anatomy of Doris, Philos. Transact. MDCCCLII, Plate 17, f. 8, is not very like nature.

three distinct commissures, which are nearly as long as the diameter of the ganglia. From the outer part of the right visceral ganglion issues a nerve nearly as long as the transverse diameter of the whole central nervous system and swelling to a rather large ganglion (gangl. penis) at the root of the penis; this ganglion contains only rather small cells and gives off three or four strong and several thinner nerves (Plate X, fig. 15). The part of the brain which gives off the nervus opticus, simulates a ganglion. The proximal ganglia olfactoria bulbiform, somewhat smaller than the buccal ganglia, but much larger than the distal ganglia olfactoria; the buccal ganglia flattened, rounded, connected by a rather short commissure; the ganglia gastro-œsophagalia rounded, having about one-fifth of the size of the last, containing one very large cell and a few smaller.

The eyes with black pigment and yellowish lens. The otocysts lying at the hinder part of the cerebral ganglia, as large as the eyes; with numerous small otokonia, which in the specimens from Kiel, were not much calcified. No trace of spicula in the leaves or other parts of the rhinophoria. The spicula of the skin were, so to speak, limited to the margins of the mantle and of the foot; in the last they were chiefly arranged perpendicularly or obliquely against the margin, except that in the foremost and hinder part of the sole some few spicula were seen scattered.

The amount of spicula in the skin seems to vary notably in the *Acanthodoris pilosa*, as seems to be the case in general in different forms of *Dorididæ*, especially, as far as hitherto known, in the *Polyceratidæ* (*Polycera*, *Ancula*). (Cf. Meyer and Moebius, Fauna der Kieler Bucht, I, 1865, pp. 52, 60.) Frey and Leuckart (Beitr. zur Kenntn. wirbellose Thiere, 1847, p. 145 described a very regular position of the spicula, but not, as it seems, in accordance with nature.

In the margin of the mantle the spicula were arranged as figured by Alder and Hanc., l. c., Part VII, Pl. 48, supplem. fig. 1, only more concentrically at the transition from the margin to the side of the body; a narrow belt of spicula crossed the back before the region of the gill. Some spicula were also seen in the tentacles. The spicula reached a notable length (at least 0.6 mm.), in old individuals they were more calcified than in younger ones. The skin was filled with unicellular glands, especially in the dorsal papillæ.[1]

The mouth-tube was wide and strong, about 1.5 mm. long; the bulbus pharyngeus in the largest individuals about 2.75 mm. long, by

---

[1] Cf. the (not very good) fig. 6 by Meyer and Moebius.

a breadth of 2.0 and a height of about 3.0 mm.; the sheath of the radula projecting backward nearly 1.0 mm.; the lip-disk sometimes surrounded by a ring of black pigment. The armature of the lip-disk entirely as shown (Pl. XII, figs. 1-4, 10-11) by me in the form from the Pacific, also the crop (Pl. XIII, fig. 2) of the bulbus.[1] The tongue in the eight specimens examined was provided with five, seven, eight, nine rows of plates, farther backwards also sixteen to twenty developed, and three younger rows; the total number amounting thus to from twenty-seven to thirty.[2] The large lateral teeth[3] yellow in the body, especially in the anterior-inferior part, with commonly five to eight denticles on the inside of the hook; sometimes, especially in the younger plates, the number of denticles rose to eleven to fifteen, sometimes the three to four outermost denticles were much larger than the rest, sometimes the denticulation was quite irregular; the height of this plate reached 0.4 mm. The outer plates (Pl. XI, fig. 1) commonly four to six, seldom seven to eight; in a series of four on the hinder part of the tongue, the outermost measured about 0.05, the next 0.09, 0.11, 0.125 mm.; they were quite colorless, compressed, with the upper side flattened, and rather erect.

The salivary glands as in the purple-colored form from the Pacific. No constant dilatation of the middle of the œsophagus (as figured, Pl. I, f. 12$g$, by Alder and Hancock), but a strong, particular one at the root as figured (l. c. Pl. I, f. 12$f$) by Alder and Hancock and by me (Gatt. nordischer Doriden, l. c. Taf. XIX, fig. 14$c$). The stomach as in the Pacific form; the intestine sometimes dilated in its first part, sometimes absolutely of the same caliber as the rest, and neither externally nor internally different from it; a little bag (biliary sac) which has been noticed by Alder and Hancock (l. c. Pl. I, fig. 12$k$), opening into the right side of this part of the intestine. The posterior visceral mass (liver) flattened and excavated on the anterior-inferior right half. The sanguineous gland whitish, convexo-concave, short and irregularly kidney-formed, with the excavation

---

[1] The first specimens of the Northern Atlantic left at my disposition being too small and too few for a thorough examination, I am obliged to refer to my examination given herewith of the form from the Pacific. Cf. moreover my figures in "Gatt. nord. Doriden," l. c. Pl. XIX, figs. 10, 11. The crop is rather well figured by Alder and Hanc. (l. c. Pl. I, f. 12$e$).

[2] According to Meyer and Moebius, the number of plates ("of the radula") is thirty-one, to Alder and Hancock, twenty-seven.

[3] Cf. my Gattungen nordischer Doriden, l. c. Taf. XIX, fig. 12.

forwards, transversely situated, with a largest diameter of 3.0 mm. The renal chamber and the syrinx as in the form from the Pacific.

The hermaphroditic gland as in this last variety, its white color contrasting with the hue of the liver. The anterior genital mass of short pyramidal form, with the point outwards, about 4.75 mm. long, the breadth and the height a little less. The ampulla of the hermaphroditic gland yellowish-white, forming a single ansa, about 4.0 mm. long, by a diameter of 0.75 mm. lying on the upper part of the back of the mucous gland. The spermatoduct yellowish, about 15.0 mm. long, constricted a little above the middle of its length; strong, sloping into the penis, which is about 1.0 mm. long. The armature of the penis entirely as in the form from the Pacific, continued backwards in the interior of the spermatoduct for a length of 6.0 mm.; the hooks rising to the height of about 0.035 mm., nearly colorless.[1] The spermatotheca (Pl. XIII, fig. 5a) spherical, of a diameter of about 2.0 mm., greenish or whitish; the spermatocysta (fig. 5b) much smaller, pyriform, yellowish; both filled with sperma. The chief duct (the vagina, fig. 5dd) very long, with several (four) longitudinal folds, which are folded again transversely; the structure seemed to resemble entirely the form from the Pacific; in the cavity was more or less sperma. The mucous gland yellow and yellowish-white; the fold of the duct with brownish-gray points, but no black pigment on the lower part of the vagina or penis.

One specimen of this typical form, with "brown mantle," and in all respects agreeing with the Atlantic, was dredged by Dall at Kyska, in June, 1873, on rocky bottom at the depth of ten fathoms.

An individual of a (in living state) "yellowish-white" variety was dredged by Dall in Popoff Strait (Shumagin Islands), on rocky bottom at a depth of six fathoms.

The animal preserved in spirits was 10.0 mm. long, by a breadth of 6.0 and a height of 4.5 mm.; the rhinophoria 1.5 mm. high, the gill 1.0 mm., the foot 3.0 mm. broad. The color yellowish-white. In the club of the rhinophoria about thirty leaves; nine branchial leaves; the anal papilla with three small protuberances; the renal pore very distinct on the right side. The genital opening very wide; the bulbus pharyngeus 2.0 mm. long; the tongue with seven rows of plates, the total number of these twenty-six (16 + 3); five external

---

[1] The armature of the penis has been first seen by H. Friele and G. Armauer Hansen (Bidr. til Kundsk. om de Norske Nudibranchiar. Christiania, Vidsk. Selsk. Forh., 1875, extras, p. 4).

plates. The diverticle of the œsophagus nearly as large as the true bulbus. The spermatoduct and the penis as usual, also the vagina; the spermatotheca of 1.6 mm. largest diameter. No trace of pigment on the vagina or penis, and the peritoneum was colorless.

Another variety of the species, with "brown mantle and yellowish-white papillæ," was dredged by Dall, in Yukon Harbor (Shumagins), in August, 1874, on sand and stones, at a depth of six to twenty fathoms.

The individual preserved in spirits was 9.0 mm. long, by a breadth of 6.5 mm., and a height of 4.5 mm.; the breadth of the foot 4.0 mm., the height of the gill 1.5 mm. The back of the animal densely brown-dotted, especially the circumference of the gill and the free area left in the middle of the gill; the dorsal papillæ all whitish; the stalk of the rhinophoria and the inferior part of the club densely dotted with brown, also, in a somewhat slighter degree, the outside of the branchial leaves. The under side of the mantle and the upper side of the margin of the foot and, in a slighter degree, the sides of the body and the sole of the foot dotted with an enormous quantity of brownish-gray points. The form as usual. The gill with nine leaves, of which the two posterior were much smaller than the others.

The central nervous system as usual; the otocysts very conspicuous under the magnifier as chalk-white points. The mouth-tube 2.0 mm. long. The bulbus pharyngeus 2.0 mm. long; the sheath of the radula projecting 2.0 mm., bent downwards. The armature of the lip-disk (Pl. XII, fig. 10, 11) very like that of the var. *albescens* (cf. Pl. XIII, fig. 4). The buccal crop as usual. The tongue with nine rows of plates; the total number of rows, twenty-five (13+3). The large lateral plates as usual; the denticulations rather long and somewhat irregular. The number of the external plates (fig. 12) reaching to six.

The salivary glands, the œsophagus with its diverticle, the pyloric part of the intestine with its bag (biliary sac), and the liver, as usual. The sanguineous gland rather large, covering, besides the central nervous system, the whole of the bulbus pharyngeus.

In the lobes of the hermaphroditic gland, masses of zoösperms. The anterior genital mass of the usual form; the ampulla of the hermaphroditic duct somewhat larger. The spermatoduct as usual; so, too, the penis, with its armature; the length of the glans about 0.5 mm. The spermatotheca and the spermatocysta as usual; also the chief duct (vagina), the cavity of the last filled with sperma. The mucous gland yellowish-white and in the centre (albuminous gland) brownish-yellow. Very scanty black pigment on the inferior part of the vagina and of the penis; the peritoneum of the back, on the contrary, very dark brown.

2. **Acanthodoris pilosa** (O. F. Müller), var. *albescens*, Pl. X, fig. 14, 15; Pl. XI, fig. 2; Pl. XII, fig. 13-16.

Color flavescente-albidus.
Hamus dentium (linguae) edentulus vel parce denticulatus.
*Habitat.* Oceanum Pacificum septentrion. (Aleutian Islands).

Two rather large specimens of this variety have been dredged by Dall, in June and July, 1873, at Kyska Harbor (Aleutians), on sand or on rocky bottom, at a depth of nine to fourteen fathoms.

According to Dall, the color of the living animal was "yellowish-white;" that of the specimens preserved in spirits was so, too, but very likely much more whitish. The length was 16.0 or 17.0 mm., by a breadth of 6.5 to 8.0 mm., and a height of 6.5 mm.; the height of the rhinophoria 2.5 to 3.0 mm., of the gill 3.0 to 4.0 mm.; the breadth of the foot 5.0 or 6.0 mm., the length of the genital opening 2.0 or 3.0 mm. The form as in the typical *D. pilosa;* the rhinophoria showed about twenty-five broad leaves in the club; there were nine branchial leaves; the anal papilla very low; the renal pore rather large.

The central nervous system as previously described. The distal olfactory ganglion small; a large (diameter, 0.4 mm.) ganglion penis (fig. 15). The eyes with rich, coal-black pigment; the otocysts visible under a lens as chalk-white points, with about one hundred and fifty otokonia.

The bulbus pharyngeus 3.5 mm. long, with the sheath of the radula projecting 1.3 to 1.5 mm.; the height of the bulbus, with the crop, 4.0 to 4.5 mm., its breadth 2.5 to 3.0 mm.

The older elements of the lip-plate (Pl. XII, figs. 13, 14) agreeing in form with those of the typical species, but oftener showing a granulated interior; the said elements reaching a length of about 0.04 mm. The diameter of the disk and mouth about 3.0 mm. The breadth of either half of the disk 0.66 mm.

The tongue showed nine or ten rows of teeth; the whole number of rows, twenty-nine (16 or 17 + 3). The large lateral teeth were as in the typical species, reaching 0.65 mm. in height (Pl. XII, fig. 15, 16), without or with only a very slight denticulation of the hook (fig. 15). The number of the outer teeth, three to five.[1]

---

[1] Cf. my Gatt. nordischer Doriden, l. c., Taf. xix, fig. 13.

The salivary glands deeply imbedded in the cavity for the œsophagus at the fore-end of the liver. The œsophagus with its rather large (1.5 mm. long) diverticle, the stomach, the intestine with its little (1.0 mm. long bag, as above. The liver 7.0 to 9.0 mm. long, 5.0 to 6.0 mm. broad, 5.0 to 6.25 mm. high, of yellowish-gray color. The sanguineous gland of irregular, oval form, of a largest diameter of 4.0 mm., by a thickness of 1.0 mm., and of grayish color. The renal syrinx about 0.75 mm. long.

The anterior genital mass 6.0 or 7.0 mm. long, 4.0 to 6.0 mm. high, and 3.0 or 4.0 mm. thick. The ampulla as usual; also the (about 40.0 mm. long) spermatoduct and the (nearly 2.0 mm. long) penis, with its armature; the hooks often set in pairs. The spermatotheca (diameter, 4.0 mm.) and the spermatocysta (diameter, 1.5 mm.) as above; the chief duct, with the vagina (about 23.0 mm. long, by a diameter of 0.4 to 1.0 mm., as usual, and also its internal cellular clothing (Pl. X, fig. 13); the yellow nucleoli somewhat brighter; the cavity nearly filled with sperma. The mucous gland as usual. No black pigment on the inferior part of the vagina or on the penis.

3. **Acanthodoris pilosa** (O. F. Müller), var. *purpurea*, Pl. XII, fig. 1-9.

Color e purpureo brunneus et flavescente-albidus.

*Habitat.* Oceanum Pacificum septentrion. Insulæ Aleutianæ (Unalashka).

Only two specimens of this species were dredged by Dall, in September, 1874, on mud and stones, at a depth of about sixty fathoms.

The color of the living animal was, according to Dall, "purple-brown and yellowish-white." The length of the animals preserved in spirits was 24.0 or 25.0 mm., by a breadth of 9.0 or 10.0 mm., and a height of 7.5 mm.; the foot 6.0 mm. broad; the height of the rhinophoria about 3.0 mm., of the branchial leaves 2.3 mm. The color of the back reddish-brown; the stalk of the rhinophoria brownish, the club yellowish; the branchial leaves yellowish-white, the last brownish at the rhachis; the under side of the mantle margin, with the sides of the body, the head and the foot, yellowish-white, dotted with brownish-gray all over, the color much more scanty on the sides of the foot and still more so on the head and on the sole of the foot.

The form was somewhat elongate. The back covered all over with pointed, rather (0.75 mm.) high, digitiform, soft papillæ and with intermixed smaller ones. The margin of the rhinophor-holes with several pointed, projecting, digitiform processes; the stout club of the rhinophoria with about twenty leaves. The branchial leaves strong, in both individuals) eight in number, the two hindermost separated by a narrow crest, which rises into a larger papilla; before this the anal papilla, covered with some papillæ, at its right side is the renal pore; on the space before it were several smaller papillæ. The under side of the free margin of the mantle (about 2.0 mm. broad) smooth. The head large, the tentacles short. The genital opening a rather large, crescentic orifice. The foot rounded behind.

The peritoneum was richly dotted on the back with brownish-red.

The central nervous system nearly quite as in *Ac. pilosa;* the proximal olfactory ganglia of oval form, true distal ones could not be detected in the root of the rhinophoria, but only a fusiform swelling of the nerve, with scattered nervous cells. The subcerebral and pedal commissures connected, the visceral isolated. The buccal ganglia larger than the olfactory, of oval form, connected by a commissure nearly as long as each ganglion; the gastro-œsophageal ganglia developed on the side of the nerve, which is a little longer than the ganglion, and in size about one-fifth of the former; the contents one very large cell, three or four smaller and several quite small ones. On the upper part of the penis the large ganglion genitale, of about the diameter of 0.3 mm., rounded, partly covered with black pigment, consisting of only rather small cells; in the first parts of the nerves given off from the ganglion, one or two rows of nervous cells of the same kind as in the ganglion.

The eyes with black pigment, yellow lens; the optic nerve rather long. As chalk-white points the otocysts were situated on the hinder part of the cerebral ganglia, where they touched the pedal ones; they were filled with solid, yellowish otokonia of about the usual form and size, but, in both respects, rather irregular. In the leaves of the rhinophoria no spicula. In the margin of the mantle and of the foot almost no spicula at all, but everywhere in the skin, especially on the back and the papilla, were an enormous quantity of large and small glandular openings. In the interstitial connective tissue were hardly any calcified cells at all.

The mouth-tube was about 2.3 mm. long, wide, with a glandular belt on the outside, not closed below; on the inside lined with a yellowish cuticula. The bulbus pharyngeus strong, about 4.0 mm. long, and the sheath of the radula projecting nearly 1.0 mm. from the posterior part of the under side, directed straight backwards or downwards; the height (through the buccal crop) 4.0 mm., the breadth 2.5 mm. The buccal crop making nearly half of the whole bulbus, and of the usual form; the walls very thick; the compressed and rather small cavity communicating through a long cleft with the anterior half of the small buccal cavity. The lip-disk (fig. 1) of rounded contour, clothed throughout its whole breadth (on each side to about 0.5 mm.) with the light, horn-yellow colored armature; the lowest part of this, as usual in the *Acanthodorides*, injured or wanting; the breadth of the belt decreasing towards the upper end, where it is interrupted in the middle line, also at the lower end. The armature (fig. 2bb, 3b, 4) composed of hooks, whose points are directed forwards (towards the opening of the mouth, nearly like, but still differing a little from those in the typical *Ac. pilosa*, reaching the height of about 0.04 mm., yellowish, with rounded, bifid or irregularly cleft points. The lancet-shaped (fig. 1a, 2a, 3a) blades at the inferior angle of the mouth as usual. The tongue with nine or ten series of plates, farther backwards thirteen to fifteen developed and three undeveloped series; the total number in this way, twenty-five to twenty-eight. The large lateral plates relatively larger than in the *Ac. pilosa*, and (fig. 5, 6) less thick in the anterior-inferior part of the body, with relatively larger hook; the denticulation of this last much weaker and much more irregular: in one specimen generally two to four denticles, sometimes only a few very insignificant ones or none at all (fig. 6); and this was the case with the other specimen, in which only some few plates showed two small denticles.[1] The outer lateral plates as in the typical form, scarcely more than from four to six.

The salivary glands whitish, rather strong at their short first part, in the rest of their length thin (fig. 7), accompanying the œsophagus to the cardia; the duct rather short (fig. 7a).

The œsophagus forming a little crop,[2] with thin walls and longitudinal folds on the inside; in the rest of its length rather thin. The stomach rather small, with the usual biliary apertures. The intestine (fig. 8a) somewhat inflated in its first part, with many rather strong folds and one particularly thick; a little over the point, where it appears on the surface of the visceral mass, on the right side, a little, scarcely pedunculated bag (fig. 8b), of the length of 1.0 to 1.25 mm., with fine, longitudinal folds; the rest of the intestine (fig. 8c) somewhat narrower; the total length of the intestine about 12.0 to 13.0

---

[1] Although very like the plates of the Atlantic form, they still bore a somewhat peculiar aspect.

[2] Cf. my Gattungen nordischer Doriden, l. c., Taf. xlx, fig. 14.

mm., by a diameter of 1.0 to 1.5 mm. The contents of the stomach and of the intestine indeterminable animal matter, mixed with an enormous quantity of different and partly very handsome forms of *Diatomaceæ*, with some *polythalamia* and some small *copepoda*, and fragments of the same.

The liver about 9–9.5 mm. long by a breadth (at the forepart) of 6.5–5.5 and a height of 6.25–6.0 mm.; the posterior half somewhat pointed, the anterior notably flattened and excavated on the right side; around the cardia the liver appeared naked (not covered by the hermaphroditic gland) of (greenish) gray color, in sections it was yellowish.

The ramifications of the aorta nearly as in the typical *Dorididæ*,[1] the root of the posterior aorta still longer and the *Art. syringis renalis* stronger and more ramified. The sanguineous gland yellowish-white, rather flattened, of irregular triangular form, lobulated, about 3.5 mm. long.

The renal chamber large; the yellowish-white renal syrinx about 0.75 mm. long, its tube somewhat more than twice as long, immediately continuous with the tube on the floor of the renal chamber.

The hermaphroditic gland easily distinguishable from the liver through its more whitish color; the secondary (ovigerous) lobes rather small; in the lobes zoösperms and large oögene cells. The anterior genital mass of plano-convex heart-shape with the point down and backwards; the length about 5.0 mm. by a breadth of 4.0 and a height of 5.0 mm. The ampulla of the very thin and white hermaphroditic duct resting on the upper posterior part of the mucous gland, yellow, short and thick (4.0 mm. long by a diameter of about 1.25 mm. forming a simple ansa. The vas deferens yellowish, strong, resting upon the upper side of the genital mass with its large coils and freely descending before its anterior margin to the penis, constricted about the (fig. 9c) middle of its total length (30.0–35.0 mm.). The penis forming the end of the spermatoduct somewhat thicker, about 2.0 mm. long, somewhat curved; its lower part hollow, the rest solid and prominent in the cavity of the former as a cylindrical glans of the length of about 0.6 mm. The glans with about ten series of yellowish hooks, which from a rather large basis raised to the height of about 0.04 mm.; the continuation of the armature reaching through the interior of the glans and of the spermatoduct nearly up to the stricture of the last, but the

---

[1] Cf. my Malacolog. Unters. (Semper, Philipp.) Tab. XLVIII, fig. 11.

number of series here smaller, about five to eight. The spermatotheca whitish, forming an oval bag of 3.0 mm. largest diameter; the spermatocysta yellowish, of 1.3–1.5 mm. largest diameter, the ducts as in the typical *Ac. pilosa*. The chief duct, too, very (about 25.0 mm.) long, rolled up in many coils, partly spiral, the diameter varying between about 0.3 and 0.75 mm.; the last fourth of the duct (vagina) with scattered black pigment, somewhat narrower and with a rather strong retractor muscle at its commencement; the interior of this duct with some few strong longitudinal folds, clothed with a cuticula, and under the same a very fine layer of round and angulated cells with a large round or oval nucleus of the diameter of about 0.4 mm. and a rather large yellow nucleolus (Pl. X, fig. 13). In the cavity of the vagina more or less sperma.[1] The mucous gland yellowish and white; the central mass (albuminous gland) yellow; the duct with scattered black pigment on the outside (also on the outside of the lower part of the penis), with the usual fold. The vestibulum genitale with black pigment on the folds, the same pigment was seen in the lowest part of the cavity of the penis and of the vagina and on the folds of the duct of the mucous gland.

A very similar animal, but "with brown mantle," was dredged by Dall in Kyska Harbor (Aleutians) in July, 1873, on sand, at a depth of nine to fourteen fathoms.

It was of large size; the length 21.0 mm., by a breadth of 11.0 and a height of 9.0 mm.; the margin of the mantle 2.0 mm. broad, the foot 6.0 mm. broad; the height of the rhinophoria and of the gill 3 mm.; the genital aperture 3.0 mm. broad. The color dirty brown on the upper side; the rhinophoria and the branchial leaves yellowish, dotted with grayish, especially on the stalk of the rhinophoria; the sole of the foot yellowish, the under side of the animal whitish; the under side everywhere with an enormous quantity of gray and black dots. The number of branchial leaves nine.

The peritoneum black-brown; the central nervous system, eyes, otocysts, as previously described. The bulbus pharyngeus of the length of 4.5 mm. by a breadth of 3.0 and a height (with the crop) of 4.75 mm.; the sheath of the radula projecting 1.25 mm.; the crop alone of the height of 2.3 mm. and 3.25 mm. broad. The lip-disk as above, the thickenings in the lowest part of the mouth 1.2 mm. long, of which nearly half freely projected. On the tongue nine rows of

---

[1] The length of the spermatoduct and the duct of the spermatotheca (vagina) was much more considerable than in the typical form.

plate, farther backwards eighteen developed and three younger rows, the total number thirty; the plates denticulated as previously mentioned, the height of the large plates rising to 0.7 mm.; the number of external plates four to five. The œsophageal diverticle of a largest diameter of about 3.0 mm. The pars pylorica of the intestine of about 4.5 mm. length, with higher folds than in the rest of the intestine, which had a length of about 15.0 mm.; the bag at the first part of the intestine 1.5 mm. long. The liver 12.0 mm. long by a breadth of 8.0 and a height of 6.0 mm. The sanguineous glands whitish, 5.0 mm. long by a breadth of 6.0 mm. and 2.0 mm. thick, convexo-concáve, the fore-end flattened (by the buccal crop), the hinder end with two transverse furrows (produced by two coils of the spermatoduct; the anterior genital mass 8.0 mm. long by a breadth of 3.5 and a height of 7.5 mm. The ampulla of the hermaphroditic duct 5.0 mm. long, whitish. The coils of the spermatoduct and of the vagina in this individual covering the upper side of the mucous gland, and ascending to the back between the pharyngeal bulbus and the liver; a coil of the former embraced the sheath of the radula. The first part of the spermatoduct 12.0 mm. long, the last of the length of about 25.0 mm; the penis about 3.5 mm. long, the armature as usual. The spermatotheca nearly spherical, of 3.5 mm. diameter; the spermatocysta yellowish, round, with a diameter of 1.5 mm.; the chief duct (vagina) 33.0 mm. long with a general diameter of 1.2 mm.; the structure of the wall as above; the last, narrower part (from the m. retractor downwards), 5.0 mm. long. The vestibulum, as well as the inferior part of the vagina and of the penis, with very scanty black pigment.

4. **Acanthodoris cærulescens**, Bgh., n. sp.   Plate XIII, fig. 6-7; Plate XIV, fig. 16.

Color paginæ superioris corporis cærulescens.
Dentes radulæ hamo per totam fere longitudinem denticulato.
*Hab.*   Mare Beringianum (Nunivak Island).

One specimen of this species was found by Dall at the north end of Nunivak Island, Bering Sea, in July, 1874, on stony bottom, at the depth of eight fathoms.

According to Dall, the color of the living animal was bluish. The animal preserved in alcohol had the length of 14.0 mm. by a height of 5.0 and a breadth of 8.0 mm.; the length of the foot was 12.5 mm. by a breadth of 6.5 mm.; the height of the rhinophoria 2.0, of the branchial leaves 1.5 mm. The color uniformly yellowish-white, with the back of a slightly bluish hue.

The form elongate-oval. The back covered all over with irregular (the greatest height reaching about 1.5 mm.), conical, rather soft and flexible papillae, in general larger than in the typical species. The margin of the rhinophor-holes thin, somewhat prominent, with two anterior strong tubercles and a posterior much smaller one; the stalk of the club rather low, the latter with about twenty-five to thirty leaves. The branchia consisting of nine to ten leaves, the adjacent border set with several strong tubercles; the branchial leaves quite isolated at their base, apparently simply pinnate. The anus prominent, before the same a small tubercle, behind it a much larger one. The margin of the mantle rather thin, on the upper side covered with a mass of smaller and larger papillae and tubercles, the under side smooth. The head broad, flat, with prominent rounded, flattened tentacula. The foot broad, rounded behind.

The central nervous system as in the typical species; the buccal ganglia rounded, the commissure between them very short. The eyes with black pigment and yellow lens. The otocysts a little smaller than the eyes, with numerous otokonia of the usual form, and reaching a length of 0.03 mm. The leaves of the rhinophoria without spicula; in the axes of the organs large, molecularly calcified cells and groups of smaller calcified cells. In the papillae of the skin of the back were no spicula at all, on their surface the usual large quantity of glandular cells; in the skin beneath the papillae cells and groups of cells as in the case of the rhinophoria.

The mouth-tube rather wide, with strong cuticula. The bulbus pharyngeus formed apparently as in the typical species: the lip-plate composed of many rows of rather low (the height rising to about 0.02 mm.), very (fig. 6) finely striated columns. The tongue with ten rows of teeth; further back, twenty-six developed and three undeveloped rows; the total number thus thirty-nine. The lateral plates large, yellow, of usual form, with a series of denticles along nearly the whole of the inner margin of the hook (fig. 16$a$). The external plates colorless, eight in number; somewhat depressed (fig. 7, 16), obliquely rising from the cuticula of the tongue (fig. 7, of nearly equal size excepting the outermost (fig. 16$b$), which is much smaller.

The salivary glands seemed of the usual form. The œsophagus and the stomach as usual. The intestine issuing from the liver at the middle of its length on the left side, rather short. The liver of the length of about 9.0 mm. by a breadth and a height of about 4.2 mm.;

the right anterior half excavated (on account of the anterior genital mass); the color brownish-gray.

The heart and the sanguineous gland as usual, also the renal chamber and the renal syrinx.

The hermaphroditic gland by its yellowish color contrasting with the liver, clothing the under side, part of the left side, and its right anterior half. The anterior genital mass rather compressed, about 6.0 mm. long by a breadth of 2.0 mm. The ampulla of the hermaphroditic duct rather short, sausage-shaped, about 2.3 mm. long, curved and whitish. The larger part of the penis was gone, but hooks were seen in the remaining part as in the typical species. The spermatotheca rather large, bag-shaped, about 3.5 mm. long; the vagina rather wide, about 10.0 mm. long. The mucous gland white, and the albuminous gland yellowish-white.[1]

This species seems very distinct from the typical one, by its color and by the different form of denticulation of the large plates of the tongue.

## POLYCERATIDÆ.

This large family, so rich in generic forms, was found represented in the northern Pacific only by two generic types, *Polycera* and *Triopha*.

### POLYCERA, Cuvier.

*Polycera*, Cuvier, (1812?), Regne-anim., 1817, ii, p. 390.[2] Regne-anim., ed. 2, iii, p. 52.
*Themisto*, Oken, Lehrb. der Zool., 1815, p. 278.
*Cufra*, Leach, Moll. Britann. Synopsis, 1852, p. 21.
*Polycera* C, Ald. and Hanc., Observ. on the genus *Polycera*, Ann. Mag. of Nat. Hist., vi, 1841, p. 337-342, Pl. IX.
*Limacia*, O. Fr. Müller, Zool. Dan., i, 1781, p. 65-68.[3]
*Phanerobranchus*, A. Frédol (Moquin-Tandon), Le monde de la mer, 1864, Pl. xii, figs. 1, 2.

---

[1] The anterior genital mass was so hardened and altered, that the nature of its different components could not be determined with certainty.

[2] According to a note of Hermannsen, under the genus *Themisto*, Oken, (Ind. Gen. Malacoz. primordia, ii, 1849, p. 572), the genus *Polycera* was established by Cuvier, 1812, [but this is probably a typographical error, since, under the genus *Polycera* itself, he indicates only the year 1817— Dall,] (cf., l. c , p. 314).

[3] *Limacia*, Hartm., Neue Alpina, i, 1821, p. 208 (*Arion*, Fér.).

Limbus frontalis digitatus vel tuberculatus. Branchia 5-7-foliata. Appendices dorsales (extrabranchiales) 1-3. Tentacula brevia, lobiformia.

Lamellæ mandibulares laterales fortes, sat applanatæ. Radula rhachide nuda; pleuris dentibus lateralibus hamatis duobus (margine lævi), interno minore, externo majore, et dentibus externus 4-8.

Prostata magna; pleuris ut in omnibus Polyceratis.

The genus *Polycera* was established by Cuvier (1812?, to receive the *Doris quadrilineata* of Müller and (in 1830) allied forms; a few years afterwards (1815), and not knowing the genus of Cuvier, Oken formed his *Themisto*, nearly identical with the *Polycera* of Cuvier.* The *Cufaca* of Leach (1852), is entirely congeneric with the genera of Cuvier and Oken, as is also very likely the *Phanerobranchus* of A. Fr. dol (Moquin-Tandon). The *Limacia* of O. Fr. Müller (1781), contains a whole series of different *Nudibranchiata*, among them the *D. quadrilineata*, and, as first-named species, the *D. verrucosa*; the name cannot therefore be employed here.

Although, through Cuvier and Alder (1841), their external characters were somewhat made known, still *Polycera*, like so many other *Nudibranchiata*, remained very superficially known, until the large monograph of Alder and Hancock,[1] that first really unveiled their external and internal structure, although Frey and Leuckart[2] had given some anatomical notices of these animals. Lately more light has been spread over the northern species of the group, through the investigations of Meyer and Moebius,[3] and of G. O. Sars.[4]

The true *Polycera* shows a form of body common to the whole family. The well-developed frontal margin is more or less curved in

---

* A careful search has failed to find any other ground for supposing that Cuvier described the genus *Polycera* in 1812, or at any date before 1817, so that the 1812 of Hermannsen is almost certainly merely a misprint. The name *Themisto*, of Oken, if congeneric, should therefore take precedence. —DALL.

[1] Alder and Hancock, Monogr. Brit. Nudibr. Moll., Part 2, 1846, fam. 1, Pl. 23; Part 4, 1848, fam. 1, Pl. 24; Part 5, 1851, fam. 1, Pl. 22; Part 6, 1854, fam. 1, Pl. 17 (anat. !); Part 7, 1855, Pl. 46 supplem. figs. 20, 21.

[2] Frey and Leuckart, Beitr. zur Kenntn. wirbellose Thiere, 1847, p. 66-70, taf. i, fig. 12, 13.

[3] Meyer and Moebius, Fauna der Kieler Bucht, i, 1865 p. 49-57, m. 2 taf. und taf. iv, A, B.

[4] G. O. Sars, Moll. reg. arct. Norv., 1878, p. 312, 313, Tab. xiv, fig. 14-16.

the middle, with its free margin tuberculated or digitate. The frontal veil is continued in a more or less tuberculated ridge, that limits the true back, and posteriorly ends in a single strong or in several smaller dorsal (branchial) appendices on the outside of and behind the region of the gill. The true back with longitudinal rows of more or less developed connected tubercles, sometimes forming low longitudinal ridges. The number of leaves in the club of the rhinophoria is not large. The gill is composed of a moderate number (five to seven) of leaves, which are either simply pinnate or composite (bi- or tripinnate). The tentacles are small, flattened or auriculate. The jaws or mandibular plates in form somewhat recall those of the *Æolidiidæ*, strong, flattened, sometimes with a peculiar superior process. The rhachis of the radula naked; on the pleuræ two large hook-formed lateral teeth, of which the outer is much larger than the inner; at the outside of the laterals are four to eight, somewhat flattened uncinæ. A large prostate gives the genital apparatus a particular feature; the armature of the penis is of the usual kind.

About the biological relations of *Polycera* very little is known, as usual among the *Nudibranchiata*. The spawn of the most common northern species is known, and a part of the developmental history has been investigated by Ray Lankester.[1]

A small number of species have been described by different authors in the course of years. Alder and Hancock (Monogr. part 7, 1855, p. 45, XVIII) established and rather well characterized two groups of *Polycera*; according to these authors Gray soon after (Guide I, 1857, p. 213) denominated these groups *Polycera* (typical) and *Palio*, which perhaps might be conserved as subgenera.

### I. POLYCERA (stricte).

Margo limbi frontalis digitatus. Folia branchialia simpliciter pinnata; appendices dorsales (branchiales) singulæ majores.

Lamellæ mandibulares processu superiori alæformi.

1. *P. quadrilineata* (O. F. Müller). M. Atlanticum; Mediterraneum.
2. *P. horrida*, Hesse. Journ. de Conchyliol., 3 S., XIII, 4, 1873, p. 345. M. Atlanticum.

---

[1] Ray Lankester, Contrib. to the Developm. hist. of Moll., Philos. Trans., MDCCCLXXV. p. 29, Pl. 10, f. 1-9.

Meyer and Moebius have, moreover, given a figure of the shell of the embryo of their *Pol. ocellata* (l. c., fig. 10).

3. *P. plebeia*, Lovén. Index Moll, 1846, p. 6.[1] M. Atlanticum.
4. *P. doriformis* (Quatref.). Phanérobranche doriforme. Moquin-Tandon (pseud. A. Frédol) Le monde de la mer., 1864, Pl. XII. fig. 1. M. Mediterraneum.
5. *P. canteriata* (Quatref.) Phanérobranche à chevrons. Moquin-Tandon (do) l. c., pl. XII, f. 2. M. Mediterraneum.

## II. PALIO, Gray.

Margo limbi frontalis tuberculatus. Folia branchialia bi- vel tripinnata ; appendices dorsales (branchiales) minores, complures.
Lamellæ mandibulares simplices (sine processu superiori).

6. *P. Lessonii* (d'Orb.). *Pol. ocellata*, A. et H. M. Atlanticum.
7. *P. pudica*, Lovén. Ind. Moll., 1846, p. 6. M. Atlanticum.
8. *P. pallida*, Bgh., n. sp. M. Pacificum.
9. *P. dubia*, Sars. Bidr. til Söedyrenés. Naturh., 1829, p. 13. Tab. 2, fig. 5, 6. Lovén, Ind. Moll., 1846, p. 6. M. Atlanticum sept.
10. *P. ? Cookii*, Angas. Journ. de Conchyl., 3 S., IV, 1, 1864, p. 58 ; Pl. V, f. 6. M. Pacificum.
11. *P. ? Capensis*. Quoy et Gaim. Voy. de l'Uranie. Zool., 1824, p. 417 ; Pl. 66 f. 4. M. Capense.[2]

**P. pallida**, Bgh., n. sp. Plate XV, fig. 11; Plate XVI. fig. 1-9.

Color flavescens. Branchia sexfoliata.
Lamellæ mandibulares fere ut in *Pol. Lessonii*, sed magis elongatæ.
Armatura lingualis fere ut in *Pol. Lessonii;* dentes externi 5.
*Hab.* Oc. Pacificum septentr.

Of this form Dall dredged a single individual in June, 1873, at Kyska Harbor (Aleutians , at the depth of ten fathoms on rocky bottom. According to Dall, the color of the living animal was "yellowish-white."

The length of the animal preserved in spirits was 7.0 mm., with a height of 4.0 and a breadth of 3.0 mm. ; the height of the branchial leaves about 1.0 mm., also that of the rhinophoria ; the breadth of the

[1] "Viridifusca, sulphureo maculata, papillis frontis 10, branchiali utrinque una postica majore ; 11 mm. Bohus." Lovén.
This, as well as the other new *Polycera* of Lovén, has not since been seen (Cf. G. O. Sars, Moll. reg. arct. Norv., 1878, p. 313).

[2] Of the three (not too naturally represented) "Polyceræ " of A. Frédol (Moquin-Tandon), the one (l. c. Pl. XII, fig. 6) seems to be the *Pol. Lessonii*, the other two (fig. 3, 4) belong undoubtedly to the genus *Thecacera*.

foot 2.0 mm. The color of the animal whitish, that of the rhinophoria and the branchial leaves more yellow; the margin of the foot white.

The form as usual. The head rounded, with a prominence on the upper lateral part; the mouth a vertical slit. The margin of the rhinophor-grooves plain. The stalk of the rhinophoria nearly as high as the club, cylindrical; the club rather flattened, with about fifteen leaves; before the rhinophoria a low transverse frontal veil with scarcely more than two prominences; the veil continued backwards as a rather indistinct prominent line on each side of the smooth rounded back; the pericardial region a little prominent; behind the middle of the length of the back, the gill with six tripinnate leaves in a slight curve; behind them the quite low anal nipple, and towards the right side the renal pore; behind the gill a little flattened space with a slight crest on each side with three papillæ. The sides of the body rather high. In the region of the anterior angles of the foot the genital papilla with the everted penis (without its recurved point, 0.75 mm. high), and below it a folded lamella, the duct of the mucous gland. The foot rather narrow, of nearly the same breadth; the rounded anterior angles somewhat prominent; a fine furrow in the anterior margin.

The intestines indistinctly appearing through the walls of the body. The peritoneum colorless, nearly without spicula.

The central nervous system (fig. 1) very depressed; the cerebral ganglia of rounded-triangular form, a little larger than the more rounded visceral (fig. 1a); the pedal ones more pyriform, a little larger than the last; the (proximal) olfactory ganglia bulbiform, not quite as large as the buccal ones, which were (fig. 1b) of rounded form, connected by a not very short commissure; the gastro-œsophageal ganglia of about one eighth of the size of the former, rounded.[1] The three inferior (subcerebral, visceral, and pedal) commissures or at least the visceral one) free.

The eyes (fig. 1) short-stalked, with black pigment and pale yellowish lens. The otocysts (fig. 1) in their usual place, very short-stalked, with about eighty otokonia of the ordinary kind. In the stalk of the rhinophoria some scattered yellowish thick spicula, of the same kind as in the skin of the back; none, on the contrary, in the leaves of the club. In the skin some scattered, yellowish, thick, straight or curved spicula, mostly of about 0.15–0.3 mm. in length, and of the usual form. In the interstitial tissue very few larger spicula.

---

[1] In the other species of *Polycera* I have examined, I never saw gastro-œsophageal ganglia, nor any in *Euplocamus* or in *Plocamopherus*.

The oral tube whitish, of about 1.0 mm. length, wide. The bulbus pharyngeus clear brownish-yellow, somewhat pyriform, with oblique flattened posterior end, in length about 1.6, by a height of nearly 1.3, and a breadth of 1.5 mm.; the sheath of the radula a little prominent downwards, and to the left from the hindermost part of the under side of the bulbus. The lip-disk clothed with a brownish-yellow cuticula, that is continued into the two mandibular plates behind the lip-disk at the entrance of the oral cavity, the form of the mandible could not be determined with certainty; a yellowish cuticula clothes the rest of the cavity. The tongue with ten rows of plates, further backwards six developed and two younger rows; the total number eighteen.[1] The rhachis (fig. 2) not very narrow. The plates yellow. The length of the first plate about 0.11, of the second 0.20, of the inmost of the external plates 0.14, of the following 0.12, 0.10, 0 08 and 0.06 mm. (all from the hinder part of the sheath). The first lateral plate (fig. 2*aa*, 5. 6) formed somewhat as in the *P. Lessonii*, the hook still smaller; the second of the same form, but larger (fig. 2*bb*, 3), the hooks much larger, especially the anterior, which is broader and excavated (fig. 7). More outwards five external plates (fig. 2*cc*), all with a crest, which is larger in the two innermost; adjoining the outermost of these plates several longitudinal folds of the lingual cuticula, which sometimes simulate one to two plates more (fig. 2).

The salivary glands whitish, elongate. The œsophagus rather wide, the stomach inclosed in the liver. The intestine appearing at the middle of the length of the liver a little to the left, at the bottom of a deep and large cavity in the upper side of the liver; the pyloric part

---

[1] According to Alder and Hancock (Monog. Part VII, 1855, Pl. 41 supplement, fig. 20, 21), the number of rows was fifteen in the *Polycera quadrilineata*, sixteen in the *P. ocellata*, thirteen in the *P. Lessonii*; Alder and Hancock saw (l. c.) four external plates in the *Pol. quadrilineata*, five in *P. ocellata*, and six in *P. Lessonii*. Meyer and Moebius saw five to seven external plates in their *Polycera ocellata*, whilst the number of rows (l. c. Pl. 50) is noted as thirteen to fifteen; in the *P. quadrilineata* they found four to five external plates and twelve to thirteen rows. In four specimens of *Pol. quadrilineata* I saw six to eight rows on the tongue, more backwards six to seven developed, and one not quite developed row; the total number of rows was fourteen to fifteen. In all specimens there were but four external plates. In four specimens of *Pol. Lessonii* I saw nine to ten rows on the tongue, more backwards eight to seven or five developed, and a single not developed row; the total number of rows was sixteen to eighteen. In all the specimens there were eight external plates.

of the intestine rather wide, its curve reaching to the bulbus pharyngeus. The liver about 5.0 mm. long by a breadth of 3.5 and a height of 3.25 mm.; the form conical, the posterior end rounded, the anterior much broader, flattened and adjoining another flattening on the inferior part of the right side of the organ; the color was yellowish.

The sanguineous gland of quadrangular form, of a diameter of about 1.5 mm., whitish.

The hermaphroditic gland with its yellowish-white lobes covering nearly the whole surface of the liver: in the lobes large o゙gene cells. The anterior genital mass of the length of about 4.0 mm. by a height of 3.0 and a breadth of 1.5 mm. The ampulla of the hermaphroditic duct resting on the inferior margin of the genital mass, whitish, straight, of the length of 3.0 mm by a diameter of about 0.5 mm. At the anterior end of the ampulla a flattened body (prostate) that freely projects before the anterior margin of the rest of the genital mass; it was of about the same length as the ampulla, but nearly twice as broad; the cavity of the organ rather large and the walls rather thin. The prostate slopes gradually into the thin but strong spermatoduct, which is about 6.0 mm. long, and terminates in the penis, which was short, conical (fig 8*a*, 9), about 0.75 mm. long, and terminated in a somewhat flexible, yellowish glans (fig. 8, 9, 14), of the length of about 0.37 mm. by a diameter at the base of about 0.09, and at the point of 0.037 mm.; through the largest part of its length it was covered with (in all about twelve) series of small chitinized crests, which did not surpass the height of about 0.0025 mm. (fig. 14); the armature only continued through a short part of the interior of the spermatoduct. The spermatotheca spherical; the spermatocysta pyriform, filled with sperma. The cordate mucous gland whitish and yellowish-white (fig. 8*b*).

This species approaches to the *Pol. Lessonii*, but seems even different in color from that and the other Atlantic forms, and also differs in the slight development of the frontal veil and of the lateral crests of the back, as well as in the number of the external plates of tongue, and[1] in the nature of the armature of the penis.

---

[1] The armature of the penis of *Polyc. quadrilineata* (hitherto the only species in which an armature has been described) as figured by Friele and Hansen (l. c. Tab. II, fig. 3 is very different from that of the Pacific species, and that difference has been confirmed by my examination of typical specimens.

**TRIOPHA,** Bergh, n. gen.

Forma corporis fere ut in Triopis, sicut quoque margo frontalis; margo dorsalis appendicibus nonnullis nodosis vel breve ramosis. Tentacula compresso-poculiformes (auriformia); rhinophoria retractilia, clavo perfoliato. Branchia quinquefoliata, foliis tripinnatis.

Os lamellis duabus fortioribus e baculis minutis compositis armatum. Lingua rhachide dentibus spuriis (4); pleuris dentibus lateralibus 3-4 (corpore processu alæformi et hamo applanato instructis) et serie dentium externorum (10-11) armatis.

Prostata?

This interesting form, that forms a link between *Polycera* and *Triopa* on one side, and the *Euplocami* on the other, approaches more nearly to the latter than to the former.

In the exterior, the *Triophæ*[1] resemble the *Triopæ*, but still differ in some points sufficiently. The appendices of the back are more composite; the tentacles seem different from those of the *Triopæ* (which have them folded lengthwise and obtuse at the end; see for comparison, Pl. XV, fig. 12); they are compressed cup-shaped or auriculate. The gill contains five leaves. Whilst the *Triopæ* want an armature of the true mouth,[2] the *Triophæ* are provided with two strong plates (composed of densely set staffs). Whilst the rhachis of the tongue in the *Triopæ* is naked, the *Triophæ* show four false plates. ("bosses" of Dall, simple thickenings of the base membrane of the radula), here; instead of the two peculiarly formed lateral plates on the pleuræ in the *Triopa*,[3] the *Triophæ* have three or four lateral plates (with a wing-like process of the body and a depressed hook); with, on the outside of these, a series of (ten to eleven) uncinal plates, nearly as in the *Triopæ*. After all, the *Triophæ* are closely allied to the *Colgæ*,[4] and essentially differ from these

---

[1] Having at first and rather superficially examined the exterior, I first regarded the animal as a *Triopa*, and called it so [s. part I, p. 128 (73), and the Plates (XIV, XV)].

[2] See for comparison Pl. XIII, fig. 19.

[3] See for comparison Pl. XIV, fig. 21, 22.

[4] The diagnosis of the *Colgæ* would be:

Forma corporis fere ut in Triopis. Vaginæ rhinophoriales calyciformes obliquæ; rhinophoria retractilia clavo perfoliato. Tentacula auriformia.

only in the armature of the tongue, which in the *Colgæ* exhibit only a single series of (false) rhachidian plates and (on each side) two lateral plates in form approaching those of *Polycera*. The nature of the prostate is unknown; the armature of the penis not differing much from that ordinary in the large group of the *Polyceratidæ*.

Although somewhat approaching to the *Euplocami* in the form of the appendices of the back, in the armature of the true mouth and of the pleuræ of the tongue, the *Triophæ* still entirely differ in the form of the tentacles, in the number of the branchial leaves and very likely in the nature of the prostate.

The *Triophæ* have hitherto been only found in the Pacific Ocean.

1. *Tr. modesta*, Bgh. n. sp. Oc. Pacificum.
2. *Tr. Carpenteri*, Stearns. Proc. of the Cal. Acad. of Sci., April 7, 1873, p. 2, fig. 2. Oc. Pacificum (California).

**Tr. modesta**, Bgh. n. sp. Pl. XIV. fig. 17-20; Pl. XV, fig. 1-10.

? *Triopa Carpenteri*, Stearns. l. c. p 2, fig. 2,

Color e flavido albescens. Appendices dorsales paucæ; folia branchialia 5.

*Hab.* Oc. Pacif. septentr.

Of this form Dall has obtained a single individual at Yukon Harbor (Shumagins), in August, 1874, at a depth of six to twenty fathoms, on a bottom of sand and stones. The color of the living animal was, according to Dall, "yellowish-white."

The animal preserved in spirits was of whitish color; the dorsal appendices, the gill and the rhinophoria more yellowish. The length of the animal 16.0 mm., by a height of 7.0 and a

Dorsum papilligerum, præsertim margo frontalis et dorsalis. Branchia paucí (4-5 foliata.

Mandibulæ triangulares, fortes. Radula fere ut in Polyceratis, dentibus lateralibus (2) et externis (7), sed præterea dentibus medianis (spuriis) instructa.

Merely one species of the genus is yet known, one of the first described *Nudibranchiata*, the *Doris lacera* of Abildgaard (Zool. Dan., IV, 1806, p. 23, Tab. CXXXVIII, fig. 3, 4), which has been found too on the coast of America (Cf. Verrill, notice of recent addit. to the Mar. Fauna of North Am., XXXVIII. Amer. Jour. of Sc. and Arts, XVI, 1878, p. 211,.

breadth of 5.5 mm.; the height of the branchial leaves 1.25, of the rhinophoria 2.0 mm.; the breadth of the foot 3.5 mm.

The form as usual. The head flattened in front, semilunar; the tentacles compressed-cup-shaped, rather short (about 1.0 mm. long), truncated at the end, longitudinally folded and open at the outer side. The frontal margin not projecting much, with many smaller and larger short digitations and crenulations; in front in the median line were two small conical papillæ before the region of the rhinophoria. The margin of the rhinophor-holes somewhat projecting, smooth: the (deeply retracted) rhinophoria with rather short stalk; the club with thirty-five to forty rather broad and thin leaves.

The back rounded over from side to side, without certain limits between it and the sides of the body. At the lateral parts of the back (on each side) five appendices; the first standing a little behind the end of the frontal margin; the next about in the middle of the space between the first and third; this last a little before the region of the gill; farther backwards were also two similar ones. The appendices were club-shaped, with simple or composite nodosities spread upon their bodies, and especially at their bases; the third was the largest, reaching the height of about 2.5 mm.; all the others a little smaller, and all of about the same size. Much smaller, conical or club-formed simple papillæ were scantily and irregularly scattered on the back. The gill consisting of five strong, tripinnate, quite separate leaves, a single anterior and two lateral pairs. The anal nipple nearly in the centre of the posteriorly open branchial circle, a blunted, nearly cylindrical prominence, about 0.5 mm. in height; at its base on the right side and a little forwards was the very distinct renal pore. The sides of the body rather high and smooth; the genital opening a short longitudinal slit lying rather forwards, with two openings at its bottom. The foot not very narrow, of nearly the same breadth throughout its whole length; the anterior border emarginated in the middle, with a fine line.

The intestines did not shine through the integuments. The peritoneum was colorless, without spicula.

The central nervous system (Pl. XV, fig. 1) flattened; the cerebro-visceral ganglia (fig. 1*a*) reniform, a little narrower at the fore-end; the pedal ones (fig. 1*bb*) rounded, scarcely larger than the visceral; the large commissure (fig. 1) as usual; small optic ganglia (fig. 1). The proximal olfactory ganglia fig. 1c) bulbiform, the n. olfactorii not very long; the distal olfactory ganglia inverse pyriform. The buccal ganglia (fig. 1dd) ovoid, connected nearly without commissure; the gastro-œsophageal ganglia small (fig. 1e), with one large cell.

The eyes (fig. 1) with coal-black pigment and yellow lens.[1] The otocysts at the usual place on the under side of the cerebro-visceral ganglionic mass, crowded with otokonia of the usual kind (fig. 1). In the leaves of the rhinophoria no spicula; in the axes and in the stalk, on the contrary, spicula of the same kind as in the skin or often larger. The skin with few and small spicula and calcified rounded cells, here and there lying in groups. The marginal dorsal appendices covered all over with above-mentioned nodosities; at their points perhaps a similar (but empty) bag as in the typical species (Cf. Pl. XIII, fig. 16, 17).

The anal tube large, 3.0 mm. long. The bulbus pharyngeus strong, of the length of 4.0 by a height of 3.0 and a breadth of 3.3 mm.; the radula-sheath projecting about 1.0 mm. from the hinder part of the under side of the bulbus. The lip-disk rather convex, with vertical oral slit (Pl. XV, fig. 2), clothed with a pale yellow cuticula, that behind the oral slit on each side is continued in a triangular, brownish-yellow lip-plate of a greatest breadth of 1.0 mm (fig. 3), narrow at the inferior end, broader at the superior, and composed of simple, somewhat curved, erect staffs (fig. 4, 5) about 0.18 mm. in height (fig. 4). The tongue broad; in the amber-yellow radula, thirteen rows of plates, further backwards in the sheath, six developed and two younger rows; the total number thus twenty-one. The three foremost rows of the tongue very incomplete, reduced to the outermost (four to five, six to seven, nine to eleven) uncinal plates. The rhachis rather broad, bearing two quadrangular thickenings of the cuticula (Pl. XV, fig. 6a) of the length of about 0.18-0.2 mm., more thickened and yellowish in the anterior margin, otherwise colorless. At the outer side of these median plates is a somewhat shorter and narrower plate (fig. 6bb), of yellowish color; in the posterior rows (Pl. XIV, fig. 20) much broader. The three succeeding plates brownish-yellow, hook-shaped, all nearly of the same form and of the same but outwardly slowly

---

[1] Alder and Hancock (l. c. part VI) also saw small optic ganglia in the *Triopa clavigera.*

decreasing size (Pl. XV, fig. 6*cd*); the fourth lateral plate, on the tongue especially, with a small hook (fig. 7*a*) that is more developed backwards, and in the four youngest rows is developed quite (Pl. XIV, fig. 17) as in the three plates mentioned. On the lateral parts of the pleuræ ten to eleven external (uncinal) plates, the four to five interior (fig. 7, 8*ab*, 10; 17*bc*) with a more developed crest, the rest (fig. 7*b*) narrower.

The salivary glands (Pl. XV, fig. 11*a*) nearly as long as the duct (fig. 11*b*); both together about 5.5 mm. long, descending along the whole back side of the bulbus pharyngeus; the gland whitish, smooth.

The œsophagus rather long (6.5 mm.), and wide especially in the posterior part (diameter 2.0 mm.), entering into the inferior part of the liver; with rather strong and numerous folds; the contents (as in the intestine) spongiary masses and different *Radiolaria* of a diameter of 0.09 mm. The intestine issuing from the liver a little before the middle of the upper side of this organ; the anteriorly proceeding part reaching the anterior margin of the liver and about 4.5 mm. long by a diameter of 1.5 mm.; the retrocessive part 7.0 mm. long by a diameter of 0.75 mm. The liver divided by a deep furrow from the right margin into two halves of nearly equal size; 6.0 mm. long by a breadth of 3.75 and a height of 3.4 mm.; the posterior extremity rounded; the anterior half of the inferior side obliquely flattened; the color yellowish-gray; the cavity of the interior rather small.

The pericardium of oval form, large, having the length of 3.5 mm. The sanguineous gland whitish, of the length of 2.5 mm. by a breadth (at the anterior end) of 2.5 mm. The renal syrinx short-pyriform; the tube of the organ strong.

The hermaphroditic gland not much developed, paler than the liver, with large oögene cells. The anterior genital mass small, about 1.5 mm. long by a height of 0.75 and a breadth of about 0.5 mm. The ampulla of the hermaphroditic duct yellowish, rather long, forming corkscrew-like windings. The spermatoduct not long, passing into the short penis. This, with its armature of very minute hooks, the spermatotheca, the spermatocysta and the vagina, as far as could be determined, as in the typical *Triopa*.[1] The gland whitish.

[1] See for comparison, Pl. XV, fig. 13.

This species may perhaps be the *Triopa Carpenteri* of Stearns; it has, like that, five branchial leaves, and does not differ much in the number of the dorsal appendices (six) or the form of the frontal margin; but the dorsal nodosities of the last species are orange-colored, and the rhinophoria, the dorsal appendices, and the branchial leaves tipped with orange. Through the great kindness of Mr. Dall I have seen a drawing of the animal of Stearns, from specimens secured after those he had described, but they do not give more details than the original description; and Stearns seems not to possess the original specimens, which very likely are lost forever. On the other side, it must be remembered that Sars (Beretn. om en i Sommeren, 1849, foretagen zoolog. Reise i Lofoten og Finmarken, 1851, p. 74) found "the young individuals of *Triopa lacera* (M.) entirely white, also on the tentacles and gills, merely the liver shines brownish through the skin."

## EXPLANATION OF THE PLATES.

An asterisk denotes that the drawing is by camera lucida, the fraction denotes the magnification.

The serial numbers of the plates (Part I, plates i–viii, Part II, plates ix–xvi) are solely referred to throughout the text. As Part II appears in another volume of the Proceedings of the Academy, the plates of Part II have been for that reason renumbered with a second set of numbers, Plate ix being Plate i, Plate x being Plate ii, etc., in the new volume. The serial numbers referred to in the text, follow the new numbers for Part II in parentheses throughout this explanation.

### Plate I (IX).

*Jorunna Johnstoni* (A. and H.).

1. *a.* stalk of the (*b*) *gangl. genitale; c. gangl. genit. secundarium*,* $\frac{200}{1}$.
2. Granules of the back, stiffened by spicula,* $\frac{200}{1}$.
3. Part of the middle of the radula, with the two innermost lateral plates; *a*, rhachis,* $\frac{350}{1}$.
4. The hook of a plate from the back,* $\frac{350}{1}$.

5. Outer part of two series of plates with 8 plates,* $\frac{350}{1}$; *aa*, outermost.
6. Outer part of another series with 3 plates,* $\frac{350}{1}$.
7. *a–b*, vagina; *c*, *gland. hastatoria*; *d*, opening of the bag of the spur; *e*, spermatoduct; *f*, penis,* $\frac{55}{1}$.
8. 9. Spermatotheca; *c*. its chief duct; *d*, *gland. hastatoria*; *b*. spermatocysta; *e*. duct to the mucous gland,* $\frac{55}{1}$.
10. *a*, Duct of the *gland. hastatoria*; *b*, the bag of the spur; *c*, opening of the bag,* $\frac{290}{1}$.
11. *a*. spermatoduct; *b*, opening of the bag at the bottom of the penis; in the interior a dart (?),* $\frac{350}{1}$.

### *Adalaria proxima* (A. and H.).

12. Tubercles of the back.
13. A part of the rhachis from above; *a*, median plates; *bb*, large lateral plates,* $\frac{75u}{1}$.
14. Part of the radula, obliquely, from the side, the hooks of the large lateral plates of both sides,* $\frac{150}{1}$.
15. Two series of (9) external plates; *a*, the innermost; *b*, the outermost,* $\frac{75u}{1}$.

### *Adalaria albopapillosa* (Dall).

16. Part of the surface of a tubercle of the back,* $\frac{350}{1}$.

### *Adalaria pacifica*, Bergh.

17. *a*, median plate; *b*, large lateral plates from the side,* $\frac{350}{1}$.

### *Lamellidoris muricata* (O. Fr. Müller).

18. The vesica fellea; *a*, its duct.

## Plate II (X).

### *Adalaria pacifica*, Bergh.

1. Median pseudo-plate (or boss), from the upper side,* $\frac{350}{1}$.
2. 2. Part of the radula, with series of (5–7) lateral plates; *a–a*. 1–2 complete rows of (15) external plates, and 1–2 incomplete rows; *bb*, innermost plates of the row; *cc*, outermost,* $\frac{350}{1}$.
3. Outer part of a row with 9 erect plates; *a*, innermost,* $\frac{150}{1}$.

### Adalaria virescens, Bergh.

4. *a*, œsophagus, with its dilatation; *b*, salivary gland; *c*, its duct.
5. *Ganglion penis*,* $\frac{200}{1}$.

### Adalaria Lovéni (A. and H.).

6. Median part of the radula from above, with (*aa*) large lateral plates; *bb*, innermost part of two rows of external plates, with 1–5 plates,* $\frac{350}{1}$.
7. Large lateral plate, from the side,* $\frac{350}{1}$.
8. Piece of the left part of the radula;* $\frac{150}{1}$ *a*, two median pseudo-plates or bosses; *b*, large lateral plates; *c*, two incomplete rows, with 6–7 plates.

### Adalaria albopapillosa (Dall).

9. *a*, (2) median pseudo-plates; *bb*, (2–3) large lateral plates of both sides.* $\frac{150}{1}$.
10. *a*, (3) median pseudo-plates; *bb*, (2–4) large lateral plates of both sides; *c*, innermost part of three (right) rows of external plates, with 3–4 plates; *d*, (left) row of 7 external plates,* $\frac{150}{1}$.
11. Four outermost plates of a row; *a*, outermost,* $\frac{150}{1}$.

### Acanthodoris pilosa (O. Fr. Müller).

12. End of the everted penis; *a*, opening,* $\frac{350}{1}$.
13. Epithelium of the vagina,* $\frac{350}{1}$.

### Acanthodoris pilosa, var. albescens (Pacifica).

14. *a*, anterior margin of the foot; *b*, edge of the tentacle.
15. *Ganglion genitale* from the penis,* $\frac{100}{1}$.

## PLATE III (XI).

### Acanthodoris pilosa (Müller).

1. Three external plates; *a*, outermost,* $\frac{350}{1}$.

### Acanthodoris pilosa, var. albescens.

2. The genital opening with its everted margin; *a*, the two foremost apertures.

### Lamellidoris bilamellata (L.) var. Pacifica.

3. Part of the branchial area with (aa) some branchial leaves; bb, some of the larger surrounding tubercles. In the centre the anal nipple, the renal pore and interbranchial tubercles.
4. The sucking crop, from the edge.
5. The half of the same, from the inside; a, stalk.
6. a, spermatotheca; b, spermatocysta; c, duct of the last; d, duct to the mucous gland; e, vagina.
7. a, two median pseudo-plates; b, a lateral plate; cc, three external plates,* $\frac{250}{1}$.
8. External plate from the side,* $\frac{250}{1}$.
9. Two of the foremost lateral plates with blunted end,* $\frac{250}{1}$.

### Lamellidoris muricata (Müller).

10. a, Median pseudo-plate shining through the left of the lateral plates, bb; c, three external plates,* $\frac{250}{1}$.
11. aa, Basal edge of three lateral plates; b, external plates,* $\frac{250}{1}$.
12. a, Glans penis; bb, praeputium; c, spermatoduct,* $\frac{100}{1}$.

### Lamellidoris varians, Bergh.

13. Lateral plate from the side,* $\frac{750}{1}$.
14. Median pseudo-plate, from above,* $\frac{350}{1}$.

### Adalaria Pacifica, Bergh.

15. Innermost part of two rows of external plates,* $\frac{250}{1}$; a, two innermost; b, the third failing (in the anterior row); c, eighth.

## Plate IV (XII).

### Acanthodoris pilosa (O. F. Müller), var. purpurea.

1. Labial disk, with (a) the lancet-formed blades projecting in the lowest part of the mouth proper.
2. The lancet-formed blades (a) with the adjoining part (b) of the armature of the mouth,* $\frac{100}{1}$.
3. a, The right lancet-formed blade; b, the adjoining part of the armature,* $\frac{350}{1}$.
4. Elements of the armature.* $\frac{250}{1}$.
5. Lateral plate, from the side,* $\frac{350}{1}$.

6. The hook of a plate, from the side.* $\frac{350}{1}$.
7. Salivary gland; *a*, duct; *b*, posterior end.
8. *a*, pars pylorica intestini; *b*, vesica fellea; *c*, intestinum descendens.
9. Part of the *vas deferens*, with its stricture,* $\frac{100}{1}$.

### Acanthodoris pilosa (M.) var. brunnea albopapillosa.

10. *ab*, Lancet-formed blades from the under side,* $\frac{100}{1}$.
11. *a*, Part of left; *b*, of right lancet-formed blade; *c*, adjoining part of the armature of the mouth,* $\frac{150}{1}$.
12. *aa*, Upper part of three lateral plates; *bb*, two series of external plates; from the sheath of the radula,* $\frac{350}{1}$.

### Acanthodoris pilosa (M.) var. albescens.

13. Elements of the armature of the mouth,* $\frac{150}{1}$.
14. Isolated element,* $\frac{250}{1}$.
15. Upper part of a lateral plate, from the outside,* $\frac{350}{1}$.
16. Upper part of a lateral plate, from the inside,* $\frac{350}{1}$.

## PLATE V (XIII).

### Lamellidoris varians, Bergh.

1. The central nervous system, obliquely, from the under side. * $\frac{55}{1}$; *a*, ganglia cerebro-visceralia; *bb*, ganglia pedalia; *c*, gangl. penis and gangl. genitale; *d*, ganglia buccalia; *ee*, ganglia gastro-œsophagalia. The eyes and the otocysts visible.

### Acanthodoris pilosa (M.), var. albescens.

2. The bulbus pharyngeus, from the side; *a*, cuticula and the lancet-formed blades; *bb*, mm. retractores bulbi; *c*, the sucking-crop; *d*, salivary gland, above this the right buccal and gastro-œsophageal ganglion; *e*, the sheath of the radula; *f*, the crop of the œsophagus; *g*, continuation of the œsophagus.
3. Lateral plates, from the outside,* $\frac{200}{1}$.
4. Part of the armature of the spermatoduct, with its hooks.* $\frac{350}{1}$.

### *Acanthodoris pilosa* (M.).

5. *a*, spermatotheca; *b*, spermatocysta; *c*, duct to the mucous gland; *dd*, duct to the vagina.

### *Acanthodoris carulescens*, Bergh.

6. Part of the armature of the mouth,* $\frac{100}{1}$.
7. External plates, from the side;* $\frac{150}{1}$ *a*, innermost.

### *Chromodoris Dalli*, Bergh.

8. The upper part of a branchial leaf,* $\frac{100}{1}$.
9. Part of the lip-plate, from above,* $\frac{150}{1}$.
10. Elements of the lip-plate,* $\frac{150}{1}$.
11. Part of the rhachis, with three (bosses or) false plates,* $\frac{150}{1}$.
12. *a*, false plate, obliquely, from the side,* $\frac{150}{1}$.
13. The 13th plate, from the side,* $\frac{150}{1}$.
14. The 9th plate, from the side,* $\frac{150}{1}$.

### *Triopa clavigera* (O. Fr. Müller).

15. Tubercles of the back.
16. Vertical section of one of the appendices of the back; *a*, bag at the point.
17. Elements of this last bag.
18. Spicula of the skin.*
19. Lowest part of the mouth, with its cuticula: *a*, the free margin,* $\frac{200}{1}$.
20. Hindermost part of the bulbus; *a*, tongue; *b*, sheath of the radula.

## PLATE VI (XIV).

### *Chromodoris Dalli*, Bergh.

1. The buccal (*a*) and gastro-œsophageal (*b*) ganglia,* $\frac{100}{1}$.
2. Part of the median portion of the radula; *a*, false plates, on each side the 2–3 innermost (lateral) plates,* $\frac{150}{1}$.
3. Outer part of two series of plates with 11 plates; *a*, outermost; *b*, eighteenth,* $\frac{150}{1}$.
4. *a*, *spermatotheca*; *b*, *spermatocysta*; *c*, duct to the vagina; *d*, duct to the mucous gland.* $\frac{5}{1}$.

### Chromodoris Californiensis, Bergh.

5. Hinder part of the body, from the under side, with 6 knots on the mantle-margin; a, foot,* $\frac{150}{1}$.
6. Upper median part of the true mouth,* $\frac{150}{1}$.
7. Part of 4 series of hooks of the lip-plate, from above,* $\frac{150}{1}$.
8–10. Elements of the same, in different positions,* $\frac{150}{1}$.
11. Three innermost plates; a, the first,* $\frac{150}{1}$.
12. One of the largest plates,* $\frac{150}{1}$.
13. Hook of 3 larger plates, obliquely, from the foreside,* $\frac{150}{1}$.
14. Four outermost plates; a, outermost,* $\frac{150}{1}$.
15. Two irregular outermost plates; a, outermost,* $\frac{150}{1}$.

### Acanthodoris cærulescens, Bergh.

16. Series of plates; a, two lateral plates; b, the outermost of the external plates,* $\frac{200}{1}$.

### Triopha modesta, Bergh.

17. Part of one of the hindermost series of plates (in the sheath), with (a) 4 lateral plates and (b, c) 2 external plates,* $\frac{200}{1}$.
18. a, second and b, third large lateral plates, from above and from the back,* $\frac{200}{1}$.
19. a, fourth; b, fifth plate (as in fig. 18 from the tongue),* $\frac{200}{1}$.
20. Outer false plate of the rhachis (from the sheath),* $\frac{350}{1}$.

### Triopa clavigera (M.).

21. a, second lateral plate; b, two external plates,* $\frac{150}{1}$.
22. First lateral plate,* $\frac{350}{1}$.

## Plate VII (XV).

### Triopha modesta, Bgh.

1. Central nervous system,* $\frac{55}{1}$; a, ganglia cerebro-visceralia; bb, pedal ganglia; c, ganglia olfactoria proximalia; dd, buccal ganglia; e, gangl. gastro-œsophagal.
2. The labial disk with the true mouth.
3. Upper commissure of the lip-plates,* $\frac{55}{1}$.
4. Elements of the lip-plate,* $\frac{350}{1}$.
5. Upper ends of two elements,* $\frac{150}{1}$.

6. Median part of a series of the teeth; *a*, (false) median plates of the rhachis; *bb*, external plate of the same; *cc*, first lateral plate; *d*, third lateral plate.* $\frac{350}{1}$.
7. Continuation of the former; *a*, fourth plate; *b*, outermost plate,* $\frac{350}{1}$.
8. Four (inner) uncinal plates; *a*, the second; *b*, the fifth,* $\frac{350}{1}$.
9. First lateral plate,* $\frac{350}{1}$.
10. Seventh and eighth external plates,* $\frac{350}{1}$.
11. Salivary gland; *a*, gland; *b*, duct,* $\frac{55}{1}$.

### *Triopa clavigera* (M.).

12. Tentacle.
13. Part of the armature of the penis.* $\frac{350}{1}$.

### *Polycera pallida*, Bergh.

14. The glans penis,* $\frac{250}{1}$.

## PLATE VIII (XVI).

### *Polycera pallida*, Bergh.

1. Central nervous system, from the upper side,* $\frac{55}{1}$; *aa*, visceral ganglia; *b*, *ganglia buccalia* and *gastro-œsophagalia*.
2. Part of the radula with two rows; *aa*, interior; *bb*, exterior lateral plates; *cc*, uncinal plates,* $\frac{350}{1}$.
3. Exterior lateral plate, from the side,* $\frac{350}{1}$.
4. Under side of the two lateral plates:* *aa* and *b*, as in fig. 2, * $\frac{350}{1}$.
5. First lateral plate, from the side,* $\frac{350}{1}$.
6. The same, from above,* $\frac{350}{1}$.
7. Hook of the second lateral plate,* $\frac{250}{1}$.
8. Genital papilla and everted penis with its glans; *b*, prominent fold of the duct of the mucous gland.
9. Glans of the penis, with the end of (*b*) the spermatoduct,* $\frac{350}{1}$; *a*, point of the glans.

### *Archidoris Montereyensis* (Cooper).

10. Large lateral plate, from the side,* $\frac{350}{1}$.
11. Outer part of two series of plates with 4 plates; *aa*, outermost,* $\frac{350}{1}$.

*Aphelodoris Antillensis*, Bergh.
(Cf. Malakozoölog. Blätter, N. S., i, 1879, p. 107–113).

12. *a*, *ganglia buccalia*, with *b*, *ganglia gastro-œsophagalia*; *c*, secondary ganglion,* $\frac{250}{1}$.
13. Median part of two series of plates; *aa*, innermost; *bb*, second plates,* $\frac{350}{1}$.
14. A large lateral plate,* $\frac{350}{1}$.
15. Outermost double plates of two series,* $\frac{350}{1}$.
16. Outer part of two series with two plates; *aa*, outermost,* $\frac{350}{1}$.
17. The sixth plate from the outer margin of the radula,* $\frac{350}{1}$.
18. Outer part of three series with 3 plates; *a*, outermost,* $\frac{350}{1}$.

*Polycera Hoibülli* (Möll.).

19. The genital papillæ, from the front.
20. The same, from the side.
21. First lateral plate, from above,* $\frac{350}{1}$.

January, 1880.

## ERRATA FOR PART I.

On account of the inability of the author to read the proofs, and from certain obscurities in the manuscript, some errors crept into the first part of this paper, and the arrangement of the paragraphs was somewhat confused by the printer.

The delicacy and beauty of the plates in their original state, having been destroyed by the printer, the present ones have been steel-surfaced, to avoid, if possible, a similar misfortune.

The specific name *Californiensis* (*Chromodoris*) was substituted in the printed text for *Calensis*, which appeared on the plate and in the manuscript under the idea that the latter was intended merely as an abbreviation.

The following list of errata has been received from the author: it is believed that the present concluding part of the paper is much less in need of such corrections.

Page 128 ( 72), line 15 : *for Triopa modesta*, B., *read Triopha modesta*, B.
" 129 ( 73), line 22 : *for* mandibuke *read* . Mandibuke.
" 130 ( 74), line  2 : *for* genus *read* penis.
" 132 ( 76), line 30 : a comma to be put before the parenthesis, and the comma after the parenthesis to be cancelled.

Page 135 ( 79), line 11 : *for* dentibus medianis denticulati *read* dentibus medianis denticulatis.
" 135 ( 79), line 18 : *for* caducous *read* not caducous.
" 135 ( 79), line 19 : a semicolon is needed before " the foot."
" 136 ( 84), line 5 : the comma after "laterales" to be cancelled.
" 136 ( 80), line 17 : a comma is needed after "1, 5"; the comma after "rhinophoria" to be cancelled.
" 138 ( 82 , line 5 : *for* Plate I. fig. 9, *read* Pl. I. fig. 9-12.
" 149 ( 84), line 39 : *for* (fig 1), one to four) *read* (pl. I. f. 11 ; pl. II. f. 1-4).
" 141 ( 85), line 1 : *for* The intestines are *read* The intestine is.
" 141 ( 85), line 3 : *for* anal papillae *read* anal papilla.
" 141 ( 85), line 34 : *for* 2 w. pl. *read* w. 2 pl.
" 141 ( 85), line 35 : *for* 2te Heft *read* 2tes Heft.
" 111 ( 85), line 41 : *for* ab *read* ob.
" 142 ( 86), line 6 : *for* denticalis *read* denticulis.
" 111 ( 88), line 16 : *for* M. retractoris *read* M. retractor.
" 145 ( 89), line 9 : *for* 3 R. J. *read* 3 R. I.
" 145 ( 89 , line 27 : *for* Dentes medianae *read* D. mediani.
" 145 ( 89), line 27 : *for* altamen *read* attamen.
" 146 ( 90), line 22 : *for* mantle *read* muzzle.
" 147 ( 91). line 11 : *for* anal *read* oral.
" 150 ( 94), line 4 : *for* Animal *read* Color animalis.
" 150 ( 94), line 3 : *before* Dendron. Dalli. B., *insert* "2."
" 152 ( 96), line 27 : *for* side, the *read* side. The.
" 153 ( 97), line 17 : *for* Dalzell *read* Dalyell.
" 153 ( 97) line 27 : *for* Tr. glauca *read* Tr. glaurae.
" 154 ( 98 , line 15 : *for* cuccnlata *read* cucullata.
" 154 ( 98), line 19 : *for* Duvancelia *read* Duvancelia.
" 155 ( 99), line 8 : *for* of the papillae *read* of the papilla.
" 156 (100), line 11 : *for* is contracted *read* was contracted.
" 156 (100), line 16 : *for* The larger mucous gland *read* The larger opening of the mucous gland.
" 156 (100), line 19 : *for* before which *read* , below which.
" 156 (100), line 38 : *for* in the hinder part *read* between the hinder parts.
" 159 (103), line 20 : *for* The cardia were wide, etc., *read* the cavity was, etc.
" 159 (103), line 26 : *for* but backward at the front and end *read* bent backward at the frontal end.
" 160 (104 , line 1 : *for* Fig. 65 a *read* 15 a.
" 161 (105), line 33 : *for* bulbus, and *read* bulbus, or.
" 161 (105 , line 38 : *for* Beitr. *read* Bidr.
" 162 (106), line 17 : *for* dentates *read* dentatis.
" 163 (107), line 33 : *for* leaves 80 *read* leaves 8.
" 163 (107), line 9 : *for* Fig. 6. 7, *read* Fig. 10, 11.
" 165 (109), line 25 : *for* Fig. 1-7 *read* Fig. 8-14.

Page 166 (110), line 19 : *for* Fig. 1 *read* Fig. 8.
" 167 (111), line  4 : *for* Fig. 2 *read* Fig. 9.
" 167 (111), line  6 : *for* Fig. 3 *read* Fig. 10.
" 167 (111), line 15 : *for* Fig. 4 *read* Fig. 11.
" 167 (111), line 16 : *for* Fig. 1 a *read* Fig. 2 a.
" 167 (111), line 16 : *for* Fig. 5 *read* Fig. 12.
" 167 (111), line 19 : *for* Fig. 4, 5. *read* Fig. 11, 12.
" 167 (111), line 23 : *for* Fig. 6, 7, 8, *read* Fig. 13, 14, 3 b.
" 168 (112), line  5 : *for* Plate XII *read* Pl. XIV.
" 168 (112), line  6 : *for* punctus *read* punctis.
" 170 (114), line  5 : *for* Fig. 13 *read* Fig. 15.
" 170 (114), line 24 : *for* latium *read* latum.
" 170 (114), line 26 : *for* minutissimus *read* minutissimis.
" 170 (114), line 33 : *for* the gills *read* the gill.
" 171 (115), line 34 : *for* Branchiæ *read* Branchia.
" 172 (116), line 17 : *for* Samso *read* Samsö.
" 173 (117), line 30 : substitute a semicolon for the period.
" 173 (117), line 31 : substitute a period for the semicolon.
" 175 (119), line 23 : *for* 1.3 *read* 13.
" 175 (119), line 23 : *for* 7-7.0 *read* 7-7.8.
" 175 (119), line 24 : *for* the light *read* the right.
" 176 (120), line  7 : *for* individual *read* individuals.
" 176 (120), line 21 : *for* leg *read* bag.
" 177 (121), line  1 : *for* branchiæ *read* branchia.
" 177 (121), line 32 : *for* of the right hand are *read* of the right hand one, is.
" 180 (124), line 10 : *for* spermatocysts *read* spermatocyst.
" 180 (124), line 33 : substitute a semicolon for the period.
" 183 (127), line  3 : *for* c *read* a.
" 183 (127), line 18 : *for* (F.) *read* (O. F. Müll.)
" 183 (127), line 21 : *for* inside *read* outside.
" 183 (127), line 23 : *for* the same *read* the same from the inside.
" 184 (128), line 13 : *for* d *read* a.
" 184 (128), line 16 : *for* b *read* a.
" 186 (130), line 12 : *for* of *read* f.
" 186 (130), line 26 : *for* 2. *read* 2, 2.
" 186 (130), line 33 : *for* e *read* c.
" 187 (131), line 27 : *for* to the twelfth *read* to b, the twelfth.
" 188 (132), line 12 : *for* cuticle *read* skin.

R. BERGH.

1–3. Adalaria pacifica, B.   4–5. Ad. virescens, B.   6–8. Ad. Loveni (A et B)
9–11. Ad. albopapillosa (D.).   12–15. Acanthodoris pilosa (M.)

1–16. *Acanthodoris pilosa* (M.) var.

1 Lamellid. varians, B.  2-5 Acanthod. pilosa M.  6.7. Ac. coerulescens, B.
R Bergh  8-14 Chromod. Dalli. B.  15-20. Triopa clavigera M.  Lovendal

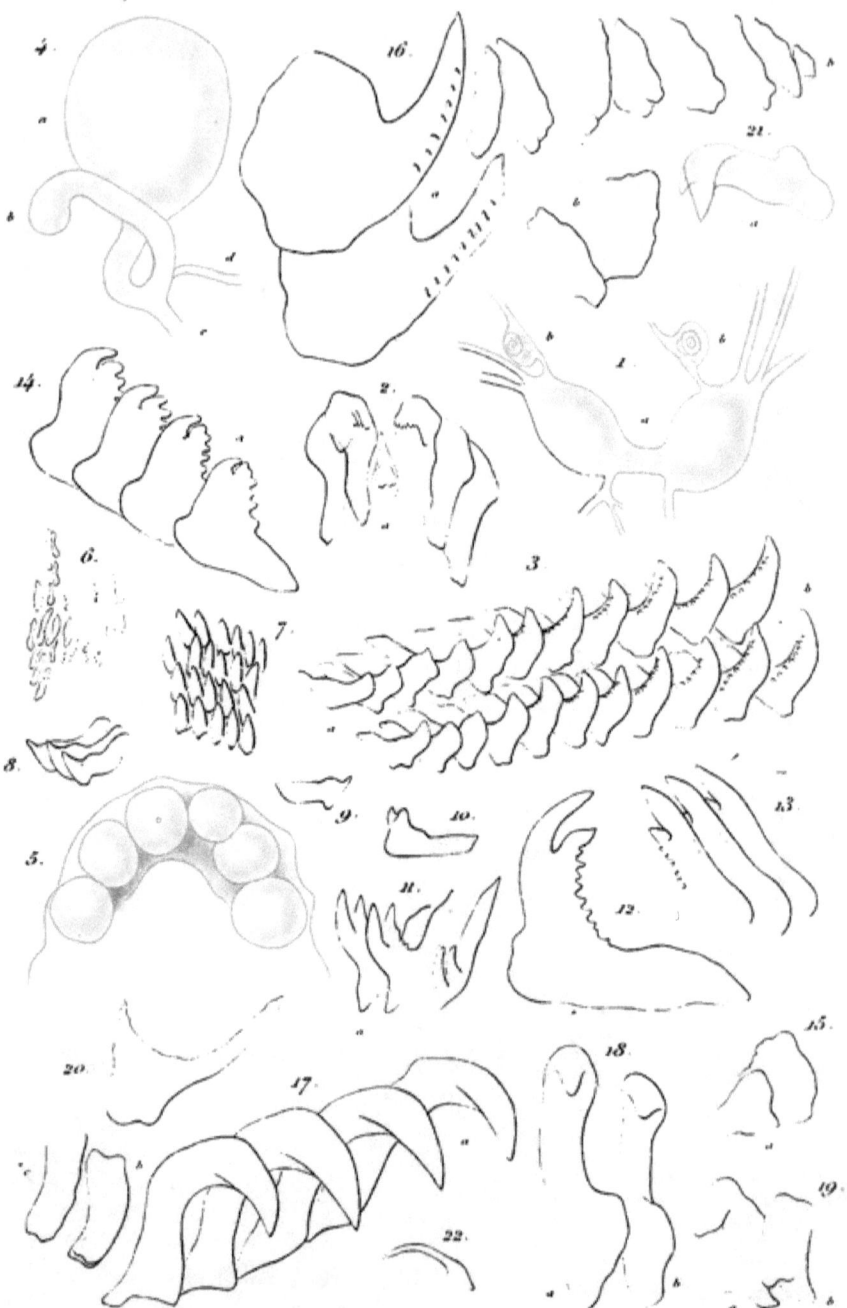

1–4. Chr. Dalli, B.   5–15. Chr. Calensis, B.   16. Ac. coerulescens, B.
17–20. Triopa modesta, B.   21–22. Tr. clavigera (M.)

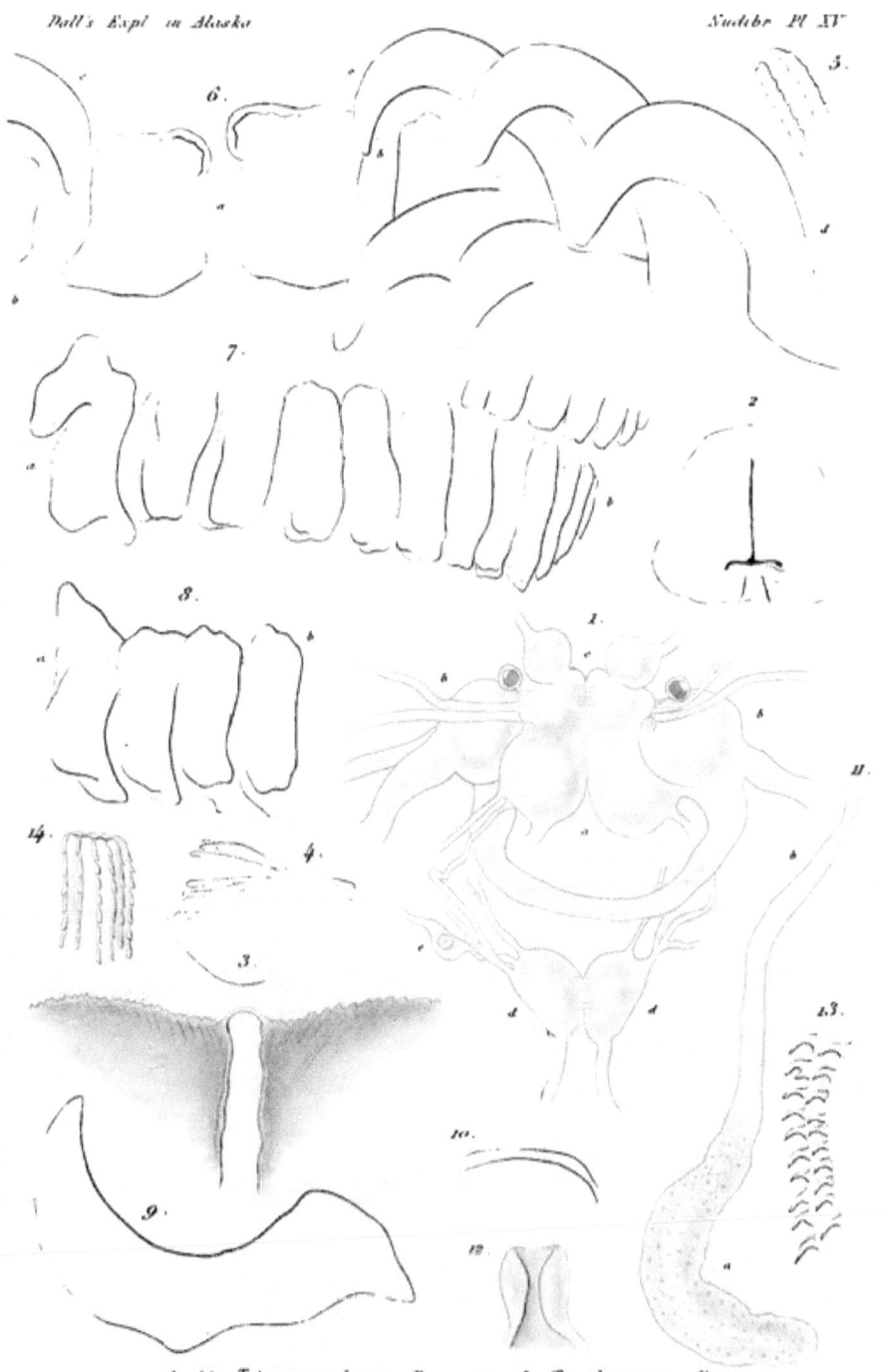

1–10. Triopa modesta. B.  11–13. Tr. clavigera. M.
14. Pol. pallida. B.

1–9. Polyc. pallida, B.
12–18. Aphelod. Antill, B.
10–11. Archid. Montereyensis (C.).
19–21. Polyc. Holbolli (M.).

www.ingramcontent.com/pod-product-compliance
Lightning Source LLC
Chambersburg PA
CBHW030404170426
43202CB00010B/1480